SpringerBriefs in Applied Sciences and Technology

More information about this series at http://www.springer.com/series/8884

Iraj Sadegh Amiri · Harith Ahmad

Optical Soliton Communication Using Ultra-Short Pulses

 Springer

Iraj Sadegh Amiri
Faculty of Science, Photonics Research
  Centre
University of Malaya
Kuala Lumpur
Malaysia

and

Laser Center, Ibnu Sina ISIR
Universiti Teknologi Malaysia
Skudai, Johor Bahru
Malaysia

Harith Ahmad
Faculty of Science, Photonics Research
  Centre
University of Malaya
Kuala Lumpur
Malaysia

ISSN 2191-530X   ISSN 2191-5318   (electronic)
SpringerBriefs in Applied Sciences and Technology
ISBN 978-981-287-557-0   ISBN 978-981-287-558-7   (eBook)
DOI 10.1007/978-981-287-558-7

Library of Congress Control Number: 2015939821

Springer Singapore Heidelberg New York Dordrecht London

Springer Science+Business Media Singapore Pte Ltd. is part of Springer Science+Business Media
(www.springer.com)

Acknowledgments

The authors would like to thank the Photonics Research Centre, Department of Physics, Faculty of Science, University of Malaya, 50603 Kuala Lumpur, Malaysia for providing the research facilities. I.S. Amiri would like to acknowledge the financial support from University Malaya/MOHE under grant number UM.C/625/1/ HIR/MOHE/SCI/29.

Contents

Abstract

In this book, we analyze the characteristics of the microring resonator (MRR) in order to perform communication using ultra-short soliton pulses. We show that rising of the nonlinear refractive indices, coupling coefficients, and radius of the single microring resonator leads to descending in input power and round trips wherein the bifurcation occurs. As a result, bifurcation or chaos behaviors are seen at lower input power of 44 W, where the nonlinear refractive index is $n_2 = 3.2 \times 10^{-20}$ m^2/W. Smallest round trips of 4770 and 5720 can be seen for the $R = 40$ μm and $\kappa = 0.1$ respectively. Here, a system of discrete optical pulse generation via a series of MRR is presented. Large bandwidth signals of optical soliton are generated and used to form continuous wavelength or frequency with large tunable channel capacity. Therefore, distinguished discrete wavelength or frequency pulses can be generated using localized spatial pulses via a networks communication system. Quantum information can be generated using a polarization control unit and a beam splitter, incorporating into the MRRs. The frequency band of 10.7 and 16 MHz and wavelengths of 206.9, 1448, 2169, and 2489 nm are localized and obtained and used for quantum codes generation applicable for networks communication. Quantum information can be performed by output signals of selected wavelengths with generated dark and bright optical soliton pulses. Using a decimal convertor system, these ultra-short signals can be converted into quantum information. Results show that multi solitons with FWHM and FSR of 10 and 600 pm can be generated respectively. The multi-optical soliton with FWHM and FSR of 325 and 880 nm are generated and can be incorporated with a time division multiple access (TDMA) system wherein the transportation of quantum information is performed.

Chapter 1
Optical Soliton Signals Propagation in Fiber Waveguides

Abstract The first observation of soliton was done by Scott Russel and the first experimental observation of soliton was occurred by using microscope objectives. In 1970s, Hasegawa primed to realize that the NLS equation was appropriate for the calculation of pulse propagation in optical fibers. Microring resonators (MRRs) shaped from nanoscale photonic waveguides. Ring resonators are not used only in optical networks, but they have recently been presented to be used as sensors, filters and biosensors. Chaotic controls have been used in a great number of optical, engineering and biological designed systems. In applications, the stored ultra-short optical signal can be used to generate optical quantum memory. The ability of ultra-short optical soliton signal to synchronize in a communication system is valid. Multi soliton generation becomes an interesting subject when it is used to enlarge the capacity of communication channels.

Keywords Soliton · Microring resonators (MRRs) · Chaotic controls · Ultra-short signal · Communication system

1.1 Optical Soliton

Optical solitons are localized as electromagnetic waves that propagate in nonlinear media resulting from a balance between nonlinearity and linear broadening due to dispersion and/or diffraction [1–4]. There are five types of nonlinear media which such as Kerr law, power law, parabolic law, dual-power law and the log law [5, 6]. In the presence of dispersive perturbation terms, the phenomena of optical soliton cooling are also observed. Initially soliton refer to the particle-like nature of solitary waves that remain intact even after common collisions [7]. The first observation of soliton was done by Scott Russel on the Edinburgh-Glasgow canal in 1834. He observed that a wave travelling through a canal without lost and major changes of its shape [8]. The first experimental observation of soliton [9] was occurred by

© The Author(s) 2015
I. Sadegh Amiri and H. Ahmad, *Optical Soliton Communication Using Ultra-Short Pulses*, SpringerBriefs in Applied Sciences and Technology,
DOI 10.1007/978-981-287-558-7_1

using microscope objectives. It was done when the mode-locked color center laser's output was coupled into the fiber, and the fiber's output into an auto correlator.

This observation appears to disagree with the nonlinear theory of Airy published in 1845, which predicted that a wave of finite amplitude cannot transmit without a change of its shape. According to his theory the wave should attenuate. The problem was solved by Joseph Boussinesq [10] in 1871. In 1876 Lord Rayleigh [11] independently, he was able to show that in a solitary wave the increase in local wave velocity associated by finite amplitude is balanced by the decrease associated with dispersion. In 1895, Korteweg de Vries [12] developed a model which can explain the unidirectional propagation of the waves of long wavelength in water with relatively shallow depth. This equation now is known as KdV. However the properties of solitons are not clearly understood until several mathematical models were introduced. The inverse scattering method was developed in the 1960s and it was able to explain the properties of soliton. The mathematical solution of soliton as KdV was found by Zabusky and Kruskal in 1964 [13].

In 1973 it was discovered that optical fibers can support dark solitons when the group-velocity dispersion (GVD) is "normal". Hasegawa and Tappert could solve and explain the non-linear Shrödinger (NLS) equation and the theory of optical soliton [14]. The first generation of spatial solitons was reported in 1974 by Ashkin and Bjorkholm [15] in a cell filled with sodium vapor. Only a decade later, Mollenauer performed the first experiment of soliton propagation in optical fibers due to the lack of adapting low loss fibers at that time [16]. Temporal dark solitons became very interesting during the 1980s [17]. During the decade of the 1990s, many other kinds of optical solitons such as spatiotemporal, Bragg, vortex, vector and quadratic solitons were discovered.

In the most recent overview of experimental observations of spatial optical solitons, some materials display large optical non-linearities when their properties are customized by the light propagation [18]. Particularly, if the non-linearity causes a change of the refractive index of the medium in which the beam can become self-trapped and propagates unchanged exclusive of any external wave guiding structure. These types of stationary self-guided beams are known as spatial optical solitons [19].

In 1970s, Hasegawa primed to realize that the NLS equation was appropriate for the calculation of pulse propagation in optical fibers, and that they should therefore support solitons. In a seminal work published in 1973 [20], he and co-author Frederick Tappert showed how the NLS equation applied to single-mode fibers, derived the essential properties of the corresponding solitons. In supporting numerical simulation they showed that the solitons were stable and robust. It is noteworthy that at the time, fibers having low loss in the region of anomalous dispersion ($\lambda > 1300$ nm) did not exist. Hasagewa and Tappert followed up almost immediately with another paper [21] describing dark solitons, i.e., sech-shaped holes in a CW background, which could exist in the presence of normal dispersion. For a number of practical reasons, however, the dark solitons have never been used for transmission.

1.2 Ring Resonators

The shape of waveguide is adjustable, thus the ring resonators can be made and used to resonate selective wavelength or can be used as filters [22–24]. MRRs shaped from nanoscale photonic waveguides. Ring resonators are employed to generate signals used for optical communication applications, where they can be integrated in a single system [25, 26]. Optical MRRs recently are interesting subject in the area of integrated optics because of their unique aspects such as compactness, low cost, tunability and easy integration on a chip with other photonic devices, having a variety of applications such as optical filter, optical switch, optical modulator, optical delay line, dispersion compensator, optical sensor and etc. [27–31].

Since ring resonators are used to support the travelling of wave resonant modes, a single ring may be applied to completely extract a particular wavelength from a signal bus. Therefore they are ideal candidates for very large-scale integrated (VLSI) photonic circuits, since they provide a wide range of optical signal processing functions while being ultra-compact [32]. MRRs have good advantages when they are used as a filter system [33–35]. The type of semiconductor microring is used widely to enhance the nonlinear optical effects which are proposed and investigated [36].

Ring resonators are not used only in optical networks, but they have recently been presented to be used as sensors, filters and biosensors [37–40]. There are many research works on the fabrication and characterization of integrated ring resonators in a variety of material systems. The fabrication and characterization of integrated AlGaAs/GaAs ring and disk resonators smaller than 10 μm was investigated in 1997, where the theoretical works based on the finite difference time domain (FDTD) analysis could be used to explain the Maxwell's equations and electromagnetic wave propagation. In 2000 Absil wrote a thesis based, on the material system AlGaAs/GaAs, where a vertical and lateral multiple coupled ring resonator configuration could be fabricated and characterized. Most of the research works on the ring resonator in the micro and nano size scale have been done since 2000. The security and high capacity of optical communication network is the major concern in the field of nano photonics.

Micro and nanoscale devices have been used widely in information technology such as telephone handsets [41–43]. One of the important parts of this device is an antenna. Using a small antenna with good performance is necessary, where to date, nano-antenna has become an interesting field in many applications such as communication and networks [44–47]. A novel nano-antenna system design is presented by Thammawongsa et al. in which photonic spins in a PANDA ring resonator are employed. These spins are generated using soliton pulse within a PANDA system. The magnetic field is introduced by using an aluminum plate coupling to the MRR, in which the spin-up and spin-down states are induced, where finally, the photonic dipoles are formed. The advantage of the proposed system is that powerful simple and compact nano-antenna can be fabricated. In addition, optical dipole can be used for further research such as dynamic dipole, dynamic torque, nano-motor, spin communicated and spin cryptography, etc. [48].

The use of data and information in optical communication is growing day by day [49–52]. Therefore, the security of data is a major concern, where there are a lot of techniques which can be used to protect the secret data or information [53–58]. Up to date, a quantum technique is recommended to provide such a requirement. A new concept of quantum cryptography using dark-bright soliton conversion behaviors within a nonlinear ring resonator (PANDA ring resonator) is presented by Amiri et al. [59–62]. In this research orthogonal soliton is established among the soliton conversion. The advantage of this research is that long distance quantum communication and high capacity quantum communication can be performed using the powerful entangle soliton. In application, the high capacity quantum communication is variable by using the multi variable entangled solitons [63].

Chaotic signals have some properties such as broadband, orthogonality and complexity aspects, which prompt researches in the areas of nonlinear science, communication technology and signal processing [64–67]. The concern in chaotic communications was due to the foreseen good properties of the chaotic signals in the fields of security systems or broadband multiple access systems. The possibility of employing chaotic signals to carry information was first studied in 1993. Encoding is the process of adding the correct transitions to the message signal in relation to the data that is to be sent over the communication system. Fiber optic sensors and micro structured fibers hold great promise for integration of multiple sensing channels. Nonlinear behavior of light inside a MRR takes place when a strong pulse of light is inserted into the ring system [68–70]. Chaotic controls have been used in a great number of optical, engineering and biological designed systems [71–74].

Optical communication is an interesting area in photonics for two decades. It is very attractive especially when it uses quantum cryptography in a network system where it was reported by Amiri et al. Quantum keys can form requires information which provides the perfect communication security [75–77]. Amiri et al. showed that quantum security could be performed via the optical-wired and wireless link. Some research works have shown that some techniques of quantum cryptography are proposed, where the systems of MRR are still complicated. Amiri et al. proposed a new quantum key distribution rule in which carrier information is encoded on continuous variables of a single photon [78–80]. In such a way, Alice randomly encodes information on either the central frequency of a narrow band single-photon pulse or the time delay of a broadband single-photon pulse. Liu and Goan studied the entanglement evolution under the influence of non-Markovian thermal environments [81]. The continuous variable systems could be two modes of electromagnetic fields or two nano-mechanical oscillators in the quantum domain, where there is no process that can be performed within a single system.

In applications, the stored ultra-short optical signal can be used to generate optical quantum memory, where the multi-soliton generation is the advantage for the systems of ring resonators. Beside improvements in efficiency and beam quality these soliton sources provide short and ultra-short pulses, leading to improved process efficiencies and new fields of laser application. The soliton pulses are so stable that its shape and velocity is preserved while travelling along the medium. The increase in communication capacity is obtained by using more available channels and large bandwidth [82, 83].

Additional information regarding these kinds of behaviors in a MRR evidently are defined by Amiri et al. [84–87]. Nonetheless, aside from the penalties of the nonlinear behaviors of light traveling within the fiber ring resonator, there are several benefits that can be employed by the communication methods in order to examine the obtained result. The chaotic behavior has been employed in digital or optical communications. The ability of chaotic carriers to synchronize in a communication system is valid. Recently, Amiri et al. have reported the successful experimental research based on generating and transmission of chaotic signals using an optical fiber communication link [29, 30, 83]. They proposed a system for chaotic signal generation and cancellation using a MRR fiber optic system, where the required signals of single bandwidth soliton pulse are recovered and manipulated using an add/drop system. Results show particular possibilities with this application. Also, effects of coupling coefficients on the bandwidth of the single soliton pulse are investigated here.

In the last few years, power system dynamics have been studied from nonlinear dynamics point of view using bifurcation theory. Nonlinear behavior of light inside a single microring resonator (SMRR) occurs when strong pulse of light is inputted into the ring system, used to many applications in signal processing and communication [88–90]. Bifurcation and chaotic signal controls with various objectives have been implemented in experimental systems and simulated theoretically used in a great number of optical, engineering and biological designed systems. Bifurcation properties of a ring system can be modified via various control methods.

One of the phenomena, known as bifurcation has been used in digital coding application. Several works have been done to show the applications of the bifurcation. Behavior of light traveling in a nonlinear ring resonator is well described by Amiri [91–93]. Laser Gaussian pulse input propagates inside the ring resonators system which is introduced by the nonlinear Kerr effect. In this book the bifurcation performance of light in a fiber microring resonator device is analyzed and characterized. Controlling of the bifurcation behavior can be implemented by controlling the round trip and input powers of the ring system via variation of the parameters. Bifurcation control is not only important in its own right, but also suggests a viable and effective approach for chaos control that can be used to generation secured codes in digital information processing, therefore, the bifurcation and chaos are usually twins [94]. Optical network is becoming a capable technology for secured and long distance communication, where optical codes are interesting tool that can offer suitable communication security and has been proposed in many research works. The MRR system can be used to generate discrete optical soliton, where, the signal bands are generated by a technique called chaotic filtering scheme. Necessary frequency bands and wavelengths are selected and can be used to generate quantum codes propagating inside network communication systems.

The entangled photon pair can be perform via the MRR system, where it can be used to generate secured key codes. Dense wavelength of optical pulses can be generated when the soliton pulse is propagating within the nonlinear MRR system and causes large bandwidth signals to be achieved, which are offered for quantum dense coding and quantum packet switching applications. Therefore, quantum encoding is implemented using the multi-entangled photon pairs [95]. Up to now, a quantum method is much useful to provide high security in optical communication network. On the other hand, the system of quantum cryptography has been widely used in many applications. Quantum key can be perform and generated using a nonlinear MRR system with appropriate parameters.

The use of quantum codes have been proposed in many research works, while the applications such as long distance, optical wireless, satellite, have been detailed. Nevertheless, a new reliable system for wireless system is needed, which has both high capacity and secure tools. Quantum codes can be performed via optical tweezers signal, generated by a MRR system in a nonlinear medium with given input power and selected parameters. By using the proposed system, different time slot entangled photons are formed randomly. Amiri et al. have proposed a technique, which can be used to communication security via the chaotic signals and up and down links of optical soliton pulses in which the use of quantum encoding of output signals is applicable [96]. They have projected the use of secured codes applicable in quantum router and network system [31]. In this book, we have used a nonlinear MRR system to form the multi wavelength, applicable for quantum codes generation used in wireless network system.

Time division multiplexing (TDM) is a type of digital multiplexing wherein two or more bit signals are transferred simultaneously as sub-channels in one communication channel. TDM can be further extended into the time division multiple access TDMA system, where several stations connected to the same physical medium, for example sharing the same frequency channel, can communicate. TDMA is a channel access method for shared medium networks in which the users receive information with different time slots. This allows multiple stations to share the same transmission medium while using only a part of its channel capacity. TDMA can be used in digital internet communications and satellite systems. Therefore, in the TDMA system, instead of having one transmitter connected to one receiver, there are multiple transmitters, where high secured signals of quantum codes along the users can be transmitted [97]. Multi soliton generation becomes an interesting subject when it is used to enlarge the capacity of communication channels. The high optical output of the ring resonator system is of benefit to long distance communication links. A Gaussian soliton can be generated in a simple system arrangement. There are many ways to achieve powerful light, for instance, using a high-power light source or reducing the radius of ring resonator.

However, there are many research works reported in both theoretical and experimental works using a common Gaussian pulse for soliton study. In practice, the intensive pulse is obtained by using erbium-doped fiber (EDF) and semiconductor amplifiers incorporated with the experimental setup. Gaussian pulse is used to form a multi soliton using a ring resonator. We propose a modified add/drop optical filter called PANDA system that consists of one centered ring resonator connected to two smaller ring resonators on the right and left sides. To form the multifunction operations of the PANDA system, for instance, to control, tune, and amplify, additional Gaussian pulse is introduced into the add port of the system. Therefore, PANDA ring resonator can be connected to an add/drop filter system in order to filter noisy and chaotic signals [83, 98, 99].

References

1. A. Afroozeh, I.S. Amiri, K. Chaudhary, J. Ali, P.P. Yupapin, Analysis of optical ring resonator, in *Advances in Laser and Optics Research* (Nova Science, New York, 2014)
2. I.S. Amiri, S.E. Alavi, S.M. Idrus, Introduction of fiber waveguide and soliton signals used to enhance the communication security, in *Soliton Coding for Secured Optical Communication Link* (Springer, New York, 2015), pp. 1–16
3. I. S. Amiri, A. Afroozeh, Introduction of soliton generation, in *Ring Resonator Systems to Perform Optical Communication Enhancement Using Soliton* (Springer, New York, 2015), pp. 1–7
4. I.S. Amiri, P. Naraei, J. Ali, Review and theory of optical soliton generation used to improve the security and high capacity of MRR and NRR passive systems. J. Comput. Theor. Nanosci. **11**(9), 1875–1886 (2014)
5. A. Afroozeh, I.S. Amiri, A. Zeinalinezhad, *Micro Ring Resonators and Applications* (LAP LAMBERT Academic Publishing, Saarbrücken, Germany, 2014)
6. I.S. Amiri, S.E. Alavi, M. Bahadoran, A. Afroozeh, S.M. Idrus, Nanometer bandwidth soliton generation and experimental transmission within nonlinear fiber optics using an add-drop filter system, J. Comput. Theor. Nanosci. (2014)
7. F. Abdullaev, J. Garnier, Optical solitons in random media. Prog. Opt. **48**, 35–106 (2005)
8. R.Y. Chiao, E. Garmire, C. Townes, Self-trapping of optical beams. Phys. Rev. Lett. **13**(15), 479 (1964)
9. L.F. Mollenauer, R.H. Stolen, J.P. Gordon, Experimental observation of picosecond pulse narrowing and solitons in optical fibers. Phys. Rev. Lett. **45**(13), 1095–1098 (1980)
10. K. Narahara, S. Nakagawa, Nonlinear traveling-wave field effect transistors for amplification of short electrical pulses. IEICE Electronics Express **7**(16), 1188–1194 (2010)
11. J. Sander, K. Hutter, On the development of the theory of the solitary wave. A historical essay. Acta Mech. **86**(1), 111–152 (1991)
12. S. Israwi, Variable depth KdV equations and generalizations to more nonlinear regimes. ESAIM Math. Model. Numer. Anal. **44**(02), 347–370 (2010)
13. G. El, R. Grimshaw, N. Smyth, Transcritical shallow-water flow past topography: finite-amplitude theory. J. Fluid Mech. **640**, 187–214 (2009)
14. X.F. Zhang, W.Q. He, P. Zhang, Controllable optical solitons in optical fiber system with distributed coefficients. Commun. Theor. Phys. **55**, 681 (2011)
15. F.W. Wise, Spatiotemporal solitons in quadratic nonlinear media. Pramana **57**(5), 1129–1138 (2001)
16. L. Mollenauer, K. Smith, Demonstration of soliton transmission over more than 4000 kmin fiber with loss periodically compensated by Raman gain. Opt. Lett. **13**(8), 675–677 (1988)
17. M. Stratmann F. Mitschke, Chains of temporal dark solitons in dispersion-managed fiber, Phys. Rev. E Stat. Nonlinear Soft Matter Phys. **72**(6 Pt 2), 066616 (2005)
18. R. Fischer, D.N. Neshev, W. Krolikowski, Y.S. Kivshar, D. Iturbe-Castillo, S. Chavez-Cerda, R. Meneghetti, D.P. Caetano, J.M. Hickmann, Observation of spatial shift in interaction of dark nonlocal solitons, pp. 136–136 (2006)

19. G.I. Stegeman, M. Segev, Optical spatial solitons and their interactions: universality and diversity. Science **286**(5444), 1518–1523 (1999)
20. A. Hasegawa, F. Tappert, Transmission of stationary nonlinear optical pulses in dispersive dielectric fibers. I. Anomalous dispersion. Appl. Phys. Lett. **23**(3), 142–144 (1973)
21. A. Hasegawa, F. Tappert, Transmission of stationary nonlinear optical pulses in dispersive dielectric fibers. II. Normal dispersion. Appl. Phys. Lett. **23**, 171 (1973)
22. A. Nikoukar, I.S. Amiri, A. Shahidinejad, A. Shojaei, J. Ali, P. Yupapin, MRR quantum dense coding for optical wireless communication system using decimal convertor, in *Computer and Communication Engineering (ICCCE) Conference* (Malaysia, 2012), pp. 770–774
23. J. Ali, M. Jalil, I.S. Amiri, A. Afroozeh, M. Kouhnavard, I. Naim, P. Yupapin, Multi-wavelength narrow pulse generation using MRR, in *ICAMN, International Conference* (Prince Hotel Kuala Lumpur, Malaysia, 2010)
24. J. Ali, I.S. Amiri, M. Jalil, M. Kouhnavard, A. Afroozeh, I. Naim, P. Yupapin, Narrow UV pulse generation using MRR and NRR system, in *ICAMN, International Conference* (Prince Hotel, Kuala Lumpur, Malaysia 2010)
25. I.S. Amiri, S.E. Alavi, S.M. Idrus, M. Kouhnavard, *Microring Resonator for Secured Optical Communication* (Amazon, USA, 2014)
26. I.S.Amiri, S.E. Alavi, S.M. Idrus, A. Afroozeh, J. Ali, *Soliton Generation by Ring Resonator for Optical Communication Application* (Nova Science Publishers, New York, 2014)
27. B.E. Little, S.T. Chu, H.A. Haus, J. Foresi, J.P. Laine, Microring resonator channel dropping filters. J. Lightwave Technol. **15**(6), 998–1005 (1997)
28. C. Tanaram, C. Teeka, R. Jomtarak, P.P. Yupapin, M.A. Jalil, I.S. Amiri, J. Ali, ASK-to-PSK generation based on nonlinear microring resonators coupled to one MZI arm. Procedia Eng. **8**, 432–435 (2011)
29. S.E. Alavi, I.S. Amiri, S.M. Idrus, A.S.M Supa'at, J. Ali, P.P. Yupapin, All optical OFDM generation for IEEE802.11a based on soliton carriers using microring resonators. IEEE Photonics J. **6**(1) (2014)
30. I.S. Amiri, S.E. Alavi, S.M. Idrus, A. Nikoukar, J. Ali, IEEE 802.15.3c WPAN standard using millimeter optical soliton pulse generated by a panda ring resonator. IEEE Photonics J. **5**(5), 7901912 (2013)
31. A. Shahidinejad, A. Nikoukar, I.S. Amiri, M. Ranjbar, A. Shojaei, J. Ali, P. Yupapin, Network system engineering by controlling the chaotic signals using silicon micro ring resonator, in *Computer and Communication Engineering (ICCCE) Conference* (Malaysia, 2012), pp. 765–769
32. N. Daldosso, L. Pavesi, Nanosilicon photonics. Laser Photonics Rev. **3**(6), 508–534 (2009)
33. I.S. Amiri, A. Afroozeh, I.N. Nawi, M.A. Jalil, A. Mohamad, J. Ali, P.P. Yupapin, Dark Soliton Array for communication security. Procedia Eng. **8**, 417–422 (2011)
34. I.S. Amiri, Multiplex and de-multiplex of generated multi optical soliton by MRRs using fiber optics transmission link. Quant. Matter (2014)
35. I.S. Amiri, A. Nikoukar, S.E. Alavi, *Soliton and Radio Over Fiber (RoF) Applications* (LAP LAMBERT Academic Publishing, Saarbrücken, 2014)
36. P. Absil, J. Hryniewicz, B. Little, P. Cho, R. Wilson, L. Joneckis, P.T. Ho, Wavelength conversion in GaAs micro-ring resonators. Opt. Lett. **25**(8), 554–556 (2000)
37. S. Saktioto, S. Daud, J. Ali, M. A. Jalil, I.S. Amiri, P.P. Yupapin, FBG simulation and experimental temperature measurement, in *ICEM* (Legend Hotel, Kuala Lumpur, Malaysia, 2010)
38. P. Sanati, A. Afroozeh, I.S. Amiri, J.Ali, L.S. Chua, Femtosecond pulse generation using microring resonators for eye nano surgery. Nanosci. Nanotechnol. Lett. **6**(3), 221–226 (2014)
39. I.S. Amiri, J. Ali, Nano particle trapping by ultra-short tweezer and wells using MRR interferometer system for spectroscopy application. Nanosci. Nanotechnol. Lett. **5**(8), 850–856 (2013)
40. I.S. Amiri, J. Ali, Picosecond soliton pulse generation using a PANDA system for solar cells fabrication. J. Comput. Theor. Nanosci. **11**(3), 693–701 (2014)

41. I.S. Amiri, S. Soltanmohammadi, A. Shahidinejad, J. Ali, Optical quantum transmitter with finesse of 30 at 800 nm central wavelength using microring resonators. Opt. Quant. Electron. **45**(10), 1095–1105 (2013)
42. I.S. Amiri, Optical soliton trapping for quantum key generation, in *The International Conference for Nano materials Synthesis and Characterization* (Malaysia, 2011)
43. A. Zeinalinezhad, S.E. Pourmand, I.S. Amiri, A. Afroozeh, Stop light generation using nano ring resonators for ROM, J. Comput. Theor. Nanosci. (2014)
44. S.E. Alavi, I.S. Amiri, A.S.M. Supa'at, *Analysis of VFSO System Integrated with BPLC* (Lap Lambert Academic Publishing, Amazon, 2014)
45. I.S. Amiri, A. Nikoukar, J. Ali, GHz frequency band soliton generation using integrated ring resonator for WiMAX optical communication, Opt. Quant. Electron. **45**(10), 1095–1105 (2013)
46. I.S. Amiri, S.E. Alavi, S.M. Idrus, RF signal generation and wireless transmission using PANDA and add/drop systems. J. Comput. Theor. Nanosci. **12**(8), 1546–1955 (2015)
47. I.S. Amiri, M. Ebrahimi, A.H. Yazdavar, S. Gorbani, S.E. Alavi, S.M. Idrus, J. Ali, Transmission of data with orthogonal frequency division multiplexing technique for communication networks using GHz frequency band soliton carrier. IET Commun. **8**(8), 1364–1373 (2014)
48. N. Thammawongsa, N. Moongfangklang, S. Mitatha, P.P. Yupapin, Novel nano-antenna system design using photonic spin in a panda ring resonator. Prog. Electromagnet. Res. **31**, 75–87 (2012)
49. I.S. Amiri, S. Ghorbani, P. Naraei, Chaotic carrier signal generation and quantum transmission along fiber optics communication using integrated ring resonators. Quant. Matter (2014)
50. I.S. Amiri, J. Ali, Data signal processing via a Manchester coding-decoding method using chaotic signals generated by a PANDA ring resonator, Chin. Opt. Lett. **11**(4), 041901(4) (2013)
51. A. Mirzaee, I.S.Amiri, *Efficient Key Management for Symmetric Cryptography System* (Amazon, USA, 2014)
52. I.S. Amiri, M. Nikmaram, A. Shahidinejad, J. Ali, Generation of potential wells used for quantum codes transmission via a TDMA network communication system. Secur. Commun. Netw. **6**(11), 1301–1309 (2013)
53. E. Fazeldehkordi, I.S. Amiri, O.A. Akanbi, *Comparative Study of Multiple Black Hole Attacks Solution Methods in MANET Using AODV Routing Protocol* (Amazon, New York 2014)
54. S.M.R.K. Soltanian, I.S. Amiri, *Detection and Defeating Distributed Denial of Service (DDoS) Attacks* (Amazon, USA, 2014)
55. I.S. Amiri, M. Ranjbar, A. Nikoukar, A. Shahidinejad, J. Ali, P. Yupapin, Multi optical Soliton generated by PANDA ring resonator for secure network communication, in *Computer and Communication Engineering (ICCCE) Conference* (Malaysia, 2012), pp. 760–764
56. I.S. Amiri, M.H. Khanmirzaei, M. Kouhnavard, P.P. Yupapin, J. Ali, Quantum Entanglement using Multi Dark Soliton Correlation for Multivariable Quantum Router, ed. by A.M. Moran in *Quantum Entanglement* (Nova Science Publisher, New York, 2012), pp. 111–122
57. I.S. Amiri, A. Afroozeh, *Ring resonator systems to perform the optical communication enhancement using soliton* (Springer, USA, 2014)
58. I.S. Amiri, S.E. Alavi, S.M. Idrus, *Soliton Coding for Secured Optical Communication Link* (Springer, New York, 2014)
59. I.S. Amiri, J. Ali, P.P. Yupapin, Enhancement of FSR and finesse using add/drop filter and PANDA ring resonator systems. Int. J. Mod. Phys. B **26**(04), 1250034 (2012)
60. I.S. Amiri, J. Ali, Generating highly dark-bright solitons by gaussian beam propagation in a PANDA ring resonator. J. Comput. Theor. Nanosci. **11**(4), 1092–1099 (2014)
61. I.S. Amiri, J. Ali, Optical quantum generation and transmission of 57–61 GHz frequency band using an optical fiber optics. J. Comput. Theor. Nanosci. **11**(10), 2130–2135 (2014)
62. I.S. Amiri, A. Afroozeh, M. Bahadoran, Simulation and analysis of multisoliton generation using a PANDA ring resonator system. Chin. Phys. Lett. **28**(10), 104205 (2011)

63. S. Tunsiri, S. Kanthavong, S. Mitatha, P. Yupapin, Optical-quantum security using dark-bright soliton conversion in a ring resonator system. Procedia Eng. **32**, 475–481 (2012)
64. I.S. Amiri, A. Nikoukar, A. Shahidinejad, J. Ali, P. Yupapin, Generation of discrete frequency and wavelength for secured computer networks system using integrated ring resonators, in *Computer and Communication Engineering (ICCCE) Conference* (Malaysia, 2012), pp. 775–778
65. J. Ali, K. Raman, A. Afroozeh, I.S. Amiri, M.A. Jalil, I.N. Nawi, P.P. Yupapin, Generation of DSA for security application, in *2nd International Science, Social Science, Engineering Energy Conference (I-SEEC 2010)* (Nakhonphanom, Thailand, 2010)
66. I.S. Amiri, S.E. Alavi, J. Ali, High capacity soliton transmission for indoor and outdoor communications using integrated ring resonators. Int. J. Commun. Syst. **28**(1), 147–160 (2013)
67. I.S. Amiri, S.E. Alavi, S.M. Idrus, Results of digital soliton pulse generation and transmission using microring resonators, in *Soliton Coding for Secured Optical Communication Link* (Springer, New York, 2015), pp. 41–56
68. I.S. Amiri, *Light Detection and Ranging Using NIR (810 nm) Laser Source* (LAP LAMBERT Academic Publishing, Germany, 2014)
69. S.E. Alavi, I.S. Amiri, S.M. Idrus, A.S.M. Supa'at, Optical amplification of tweezers and bright soliton using an interferometer ring resonator system. J. Comput. Theor. Nanosci. (2014)
70. I.S. Amiri, S.E. Alavi, S.M. Idrus', Solitonic pulse generation and characterization by integrated ring resonators, in *5th International Conference on Photonics 2014 (ICP2014)* (Kuala Lumpur, Malaysia, 2014)
71. S. Saktioto, J. Ali, M. Hamdi, I.S. Amiri, Calculation and prediction of blood plasma glucose concentration, in *ICAMN, International Conference* (Prince Hotel, Kuala Lumpur, Malaysia, 2010)
72. I.S. Amiri, S.E. Alavi, A. Shahidinejad, A. Nikoukar, T. Anwar, A.S.M. Supa'at, S.M. Idrus, N.K. Yen, Characterization of ultra-short soliton generation using MRRs, in *The 2014 Third ICT International Student Project Conference (ICT-ISPC2014)* (Thailand, 2014)
73. I.S. Amiri, A. Afroozeh, Integrated ring resonator systems, in *Ring Resonator Systems to Perform Optical Communication Enhancement Using Soliton* (Springer, 2015), pp. 37–47
74. I.S. Amiri, A. Afroozeh, Soliton generation based optical communication, in *Ring Resonator Systems to Perform Optical Communication Enhancement Using Soliton* (Springer, New York, 2015), pp. 49–68
75. I.S. Amiri, J. Ali, Femtosecond optical quantum memory generation using optical bright soliton. J. Computat. Theor. Nanosci. **11**(6), 1480–1485 (2014)
76. S.E. Alavi, I.S. Amiri, S.M. Idrus, A.S.M. Supa'at, Generation and wired/wireless transmission of IEEE802.16 m signal using solitons generated by microring resonator, Opt. Quant. Electron. **50**(8), 622–628 (2014)
77. J. Ali, I.S. Amiri, A. Afroozeh, M. Kouhnavard, M. Jalil, P. Yupapin, Simultaneous dark and bright soliton trapping using nonlinear MRR and NRR, in *ICAMN, International Conference* (Prince Hotel, Kuala Lumpur, Malaysia, 2010)
78. I.S. Amiri, K. Raman, A. Afroozeh, M.A. Jalil, I.N. Nawi, J. Ali, P.P. Yupapin, Generation of DSA for security application. Procedia Eng. **8**, 360–365 (2011)
79. I.S. Amiri, A. Afroozeh, Mathematics of soliton transmission in optical fiber, in *Ring Resonator Systems to Perform Optical Communication Enhancement Using Soliton* (Springer, Berlin, 2015), pp. 9–35
80. A. Nikoukar, I.S. Amiri, S.E. Alavi, A. Shahidinejad, T. Anwar, A.S.M. Supa'at, S.M. Idrus, L.Y. Teng', Theoretical and simulation analysis of the add/drop filter ring resonator based on the Z-transform method theory, in *The 2014 Third ICT International Student Project Conference (ICT-ISPC2014)* (Thailand, 2014)
81. K.-L. Liu, H.-S. Goan, Non-Markovian entanglement dynamics of quantum continuous variable systems in thermal environments. Phys. Rev. A **76**(2), 022312 (2007)

82. I.S. Amiri, FWHM measurement of localized optical soliton, in *The International Conference for Nano materials Synthesis and Characterization* (Malaysia, 2011)

83. I.S. Amiri, S.E. Alavi, S.M. Idrus, A.S.M. Supa'at, J. Ali, P.P. Yupapin, W-Band OFDM transmission for radio-over-fiber link using solitonic millimeter wave generated by MRR. IEEE J. Quant. Electron. **50**(8), 622–628 (2014)

84. I.S. Amiri, S.E. Alavi, F.J. Rahim, S.M. Idrus, Analytical treatment of the ring resonator passive systems and bandwidth characterization using directional coupling coefficients. J. Comput. Theor. Nanosci. (2014)

85. I.S. Amiri, R. Ahsan, A. Shahidinejad, J. Ali, P.P. Yupapin, Characterisation of bifurcation and chaos in silicon microring resonator. IET Commun. **6**(16), 2671–2675 (2012)

86. I.S. Amiri, P. Naraei, Optical transmission characteristics of an optical add-drop interferometer system. Quant. Matter (2014)

87. I.S. Amiri, S.E. Alavi, S.M. Idrus, Theoretical background of microring resonator systems and soliton communication, in *Soliton Coding for Secured Optical Communication Link* (Springer, New York, 2015), pp. 17–39

88. I.S. Amiri, F.J. Rahim, A.S. Arif, S. Ghorbani, P. Naraei, D. Forsyth, J. Ali, Single soliton bandwidth generation and manipulation by microring resonator. Life Sci. J. **10**(12s), 904–910 (2014)

89. I.S. Amiri, B. Barati, P. Sanati, A. Hosseinnia, H.R.M. Khosravi, S. Pourmehdi, A. Emami, J. Ali, Optical stretcher of biological cells using sub-nanometer optical tweezers generated by an add/drop microring resonator system. Nanosci. Nanotechnol. Lett. **6**(2), 111–117 (2014)

90. Y.S. Neo, S.M. Idrus, M.F. Rahmat, S.E. Alavi, I.S. Amiri, Adaptive control for laser transmitter feedforward linearization system, IEEE Photonics J. **6**(4), 1–10 (2014)

91. N.J. Ridha, F.K. Mohamad, I.S. Amiri, Saktioto, J. Ali, P.P. Yupapin, Controlling center wavelength and free spectrum range by MRR radii, in *International Conference on Experimental Mechanics (ICEM)* (Kuala Lumpur, Malaysia, 2010)

92. A. Afroozeh, I.S. Amiri, A. Zeinalinezhad, S.E. Pourmand, H. Ahmad, Comparison of control light using kramers-kronig method by three waveguides. J. Comput. Theor. Nanosci. **12**, 1–6 (2015)

93. I.S. Amiri, A. Shahidinejad, Generating of 57–61 GHz frequency band using a panda ring resonator. Quant. Matter **37**(38), 2 (2014)

94. P. Yupapin, P. Saeung, W. Suwancharoen, Coupler-loss and coupling-coefficient dependence of bistability and instability in a fiber ring resonator: nonlinear behaviors. J. Nonlinear Opt. Phys. Mater. **16**(01), 111–118 (2007)

95. I.S. Amiri, A. Nikoukar, Quantum information generation using optical potential well, in *Network Technologies & Communications (NTC) Conference* (Singapore, 2010–2011)

96. I.S. Amiri, A. Afroozeh, M. Bahadoran, J. Ali, P.P. Yupapin, Up and down link of soliton for network communication, in *National Science Postgraduate Conference, NSPC* (Universiti Teknologi Malaysia, Malaysia, 2011)

97. S. Kawanishi, Ultrahigh-speed optical time-division-multiplexed transmission technology based on optical signal processing: feature issue on fundamental challenges in ultrahigh-capacity optical fiber communication systems. IEEE J. Quant. Electron. **34**(11), 2064–2079 (1998)

98. I.S. Amiri, A. Afroozeh, Spatial and temporal soliton pulse generation by transmission of chaotic signals using fiber optic link, in *Advances in Laser and Optics Research*, vol. 11 (Nova Science Publisher, New York, 2014)

99. S.E. Alavi, I.S.Amiri, A.S.M. Supa'at, S.M. Idrus, Indoor data transmission over ubiquitous infrastructure of powerline cables and LED lighting, J. Comput. Theor. Nanosci. **12**, 599–604 (2014)

Chapter 2
MRR Systems and Soliton Communication

Abstract The nonlinear Schrödinger equation (NLSE) is an appropriate equation for describing the propagation of light in optical fibers. The optical solitons are very stable against perturbations, where it is property of a fundamental soliton makes it an ideal candidate for optical communications. To have the secured communication, the performance of resonators should be considered in terms of resonance width, the free spectral range, the finesse, and the quality factor. When an optical soliton pulse input into the nonlinear microring resonator (MRR), the large optical bandwidth of the output signals can be generated, where the nonlinear behavior of self-phase modulation (SPM) keeps the large output power. High capacity of optical pulses can be obtained if the full width at half maximum (FWHM) of these pulses are very small, thus the intensity build up can be performed inside the micro and nanoring systems.

Keyword Nonlinear Schrödinger equation (NLSE) · Optical fibers · Microring resonators (MRRs) · Soliton communication · Self-phase modulation (SPM)

2.1 Evaluation of Soliton

The nonlinear Schrödinger equation (NLSE) is an appropriate equation for describing the propagation of light in optical fibers using normalization parameter such as: the normalized time T_0, the dispersion length L_D and peak power of the pulse P_0 the nonlinear Schrödinger equation in the terms of normalized coordinates can be written as [1]:

$$i\left(\frac{\partial u}{\partial z}\right) - \frac{5}{2}\left(\frac{\partial^2 u}{\partial t^2}\right) + N^2 |u|^2 u + i\left(\frac{\alpha}{2}\right)u = 0 \qquad (2.1)$$

where u (z, t) is pulse envelope function, z is propagation distance along the fiber, N is an integer designating the order of soliton and α is the coefficient of energy gain per unit length, and with negative value it represents energy loss. Here, s is −1 for

© The Author(s) 2015

I. Sadegh Amiri and H. Ahmad, *Optical Soliton Communication Using Ultra-Short Pulses*, SpringerBriefs in Applied Sciences and Technology, DOI 10.1007/978-981-287-558-7_2

Fig. 2.1 Evolution of soliton in normal dispersion regime

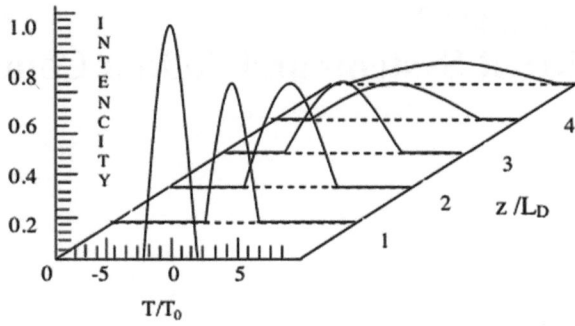

Fig. 2.2 Evolution of soliton in anomalous dispersion regime

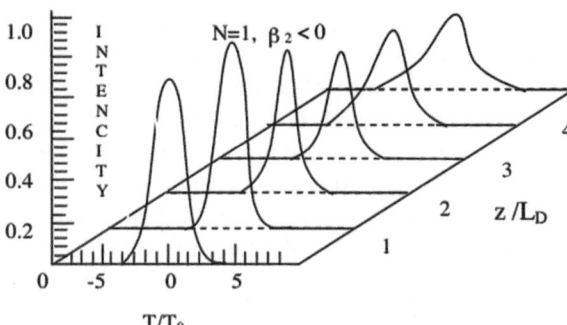

negative β_2 (anomalous GVD-Bright soliton) and +1 for positive β_2 (normal GVD-Dark soliton) as shown in Figs. 2.1 and 2.2 [2],

$$N_2 = \frac{L_D}{L_{NL}} = \frac{\gamma P_0 T_0^2}{|\beta_2|^2} \tag{2.2}$$

with nonlinear parameter γ and nonlinear length L_{NL}.

It is apparent that SPM dominates for N > 1 while for N < 1 dispersion effects dominates. For N ≈ 1 both SPM and GVD cooperate in such a way that the SPM-induced chirp is just right to cancel the GVD induced broadening of the pulse. The optical pulse would then propagate undistorted in the form of soliton. By integrating the NLS, the solution for the fundamental soliton can be written as [3, 4]

$$u(z,t) = \sec h(t) \exp(iz/2) \tag{2.3}$$

where, sec h(t) is hyperbolic scent function. Since the phase term exp (iz/2) has no influence on the shape of the pulse, the soliton is independent of z and hence is non dispersive in time domain. It is property of a fundamental soliton makes it an ideal candidate for optical communications. Optical solitons are very stable against perturbations [5]; therefore they can be created even when the pulse shape and peak

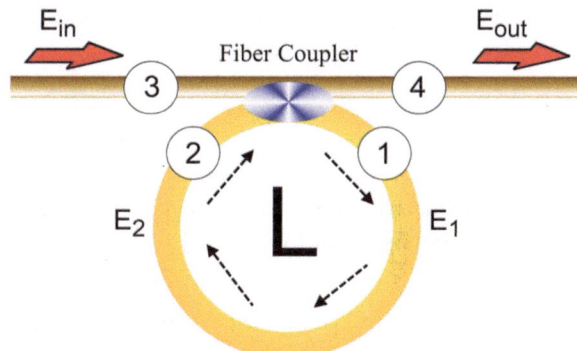

Fig. 2.3 Schematic diagram for a ring resonator coupled to a single waveguide

power deviates from ideal conditions (values corresponding to N = 1). Soliton can be used for Secured Communication. To have the secured communication, the performance of resonators should be considered in terms of resonance width, the free spectral range, the finesse, and the quality factor. A fiber optic ring resonator consists of a waveguide in a closed loop which is coupled to one or more input/output (or bus) waveguides. A simple MRR is shown in Fig. 2.3.

2.2 MRR Used to Generate Chaotic Signals

Ring resonator provides traveling wave procedure, unlike the standing wave characteristic of Fabry-Perot resonators (F-P) [6–8]. Ring resonator can be considered as an interferometer device, which resonates for light whose phase change is an integer multiple of 2π after each trip around the ring. Part of light that does not contribute this resonant condition will be transmitted through the bus waveguide. Signal loss occurs when light is transmitted through the fiber, especially over long distances such as undersea cables. The expression for the resonant wavelengths of the ring is very similar to that of the F-P and is given by

$$\lambda_r = \frac{2\pi R n_{eff}}{m} \tag{2.4}$$

where, R is the ring radius constructed with circular waveguide and m is an integer. In this situation the device will act as a phase filter where all wavelengths are transmitted and the resonant wavelengths, having also traversed the ring, acquires a phase change. To capture or separate the resonant wavelengths from the rest, an additional waveguide as an output bus, can be positioned on the opposite side of the ring. In this case the ring resonator is known as an add/drop filter system [9, 10]. The key performance parameters of the ring resonator include the free spectral range (FSR), the extinction ratio (ER), and the finesse [11]. The expression for the FSR of a ring resonator is given by

$$\Delta\lambda = \frac{\lambda_r^2}{2\pi R n_g} \tag{2.5}$$

The nonlinearity of the fiber ring is of the Kerr-type, wherein the nonlinear refractive index is given by [12, 13]

$$n = n_0 + n_2 I = n_0 + \left(\frac{n_2}{A_{\mathit{eff}}}\right) P, \tag{2.6}$$

where n_0 and n_2 are the linear and nonlinear refractive indices, while I and P are the optical intensity and optical field power, respectively. Here, the fiber coupler is considered as a point device and is reciprocal. The linear and nonlinear phase shifts of the ring resonator can be expressed by $\phi_0 = kLn_0$ and $\varphi_{NL} = kLn_2|E_1|^2$, where $k = 2\pi/\lambda$ is a wave number, and $L = 2\pi R$ is the circumference of the ring resonator, where R is the radius of the ring resonator. Mathematically, the subsequence equations of the round-trip within the system is given by [14, 15].

$$E_{n+1} = j\sqrt{(1-\gamma)\kappa}E_{in} + \sqrt{(1-\gamma)(1-\kappa)}xE_n\exp(-j(\phi_0 + \phi_{NL})) \tag{2.7}$$

Here, the subscript n denotes the number of round-trips inside the system. This equation has to be satisfied with boundary conditions appropriate for ring. The transmission around the single ring resonator is represented by

$$z^{-1} = \exp(-\alpha L/2 - jk_n L) \tag{2.8}$$

where k_n is the propagation constant and $\alpha L/2$ is the ring loss (round-trip loss), which includes propagation loss, losses resulting from transitions in the curvature, and bending losses. The value of α (unit length^{-1}) depends on the properties of the material and the waveguide used, and it is referred to as the intensity attenuation coefficient, where L is the circumference of the ring resonator. In order to describe this, we consider a ring resonator connected to a single coupler that extracts light from the ring into the output waveguides [16, 17].

When an input electric field, E_i is coupled to the ring waveguide through an external bus waveguide, a positive feedback is induced and the field inside the ring resonator, E_2 starts to build up. The feedback mechanism will be induced by the ring waveguide [18], therefore does not need any further requirements such as Bragg gratings, mirrors, or distributed feedback waveguides with difficult fabrication process. Due to on-resonant certain wavelength of the input signals inside the ring waveguide, frequency selectivity is obtained. The inserted and transmitted electric fields into the ring resonator are expressed by

$$E_1 = (1 - \gamma)^{\frac{1}{2}} \left[jE_i \sqrt{\kappa} + E_2 \sqrt{1 - \kappa} \right] \tag{2.9}$$

$$E_2 = E_1 \exp\left(-\frac{\alpha}{2} L - jk_n L \right) \tag{2.10}$$

where $k_n = \frac{2\pi \cdot n_{eff}}{\lambda}$ and γ denotes the intensity insertion loss coefficient of the directional coupler and n_{eff} is the effective refractive index. Therefore, the refractive index n quantifies the increase in the wave number (phase change per unit length) caused by the medium [19, 20]. Here, the effective refractive index n_{eff} has the similar meaning with light propagation in a waveguide, where it depends not only on the wavelength but also on the mode, in which the light propagates. The ratio of the output and input powers which is E_t/E_i can be calculated as [21, 22].

$$\frac{E_t}{E_i} = (1 - \gamma)^{\frac{1}{2}} \cdot \left[\frac{\sqrt{1 - \kappa} - (1 - \gamma)^{\frac{1}{2}} \cdot \exp\left(-\frac{\alpha}{2} L - jk_n L \right)}{1 - (1 - \gamma)^{\frac{1}{2}} \cdot \sqrt{1 - \kappa} \cdot \exp\left(-\frac{\alpha}{2} L - jk_n L \right)} \right] \tag{2.11}$$

In the following new parameter will be used for simplifying:

$$D = (1 - \gamma)^{\frac{1}{2}}, \ x = D \cdot \exp\left(-\frac{\alpha}{2} \cdot L \right), \ y = \sqrt{1 - \kappa}, \ \phi = k_n L$$

Intensity relation to the output port is given by:

$$T = \frac{I_t}{I_i}(\varphi) = \left| \frac{E_t}{E_i} \right|^2 = D^2 \cdot \left[1 - \frac{(1 - x^2) \cdot (1 - y^2)}{(1 - xy)^2 + 4xy \cdot \sin^2\left(\frac{\varphi}{2} \right)} \right] \tag{2.12}$$

Maximum and minimum transmission can be calculated when $\sin^2\left(\frac{\varphi}{2} \right)$ is "1" and "0" respectively. Therefore;

$$T_{\max} = D^2 \cdot \frac{(x + y)^2}{(1 + x \cdot y)^2} \tag{2.13}$$

$$T_{\min} = D^2 \cdot \frac{(x - y)^2}{(1 - x \cdot y)^2} \tag{2.14}$$

The minimum transmission, $T_{\min}$ occurs at the resonant point when the circumference of the ring L, is an integer number of the guide wavelength, which is given by

$$\phi = k_n \cdot L = 2 m\pi, \text{m} = \text{integer},$$
$$m \cdot \lambda_m = n \cdot L \tag{2.15}$$

Here, m is the mode number, λ_m is the resonant mode wavelength. The on-off ratio for the single ring resonator is defined as the ratio of the on-resonance intensity to the off-resonance intensity which is maximum at $T_{min} = 0$. Therefore $x = y$ and

$$\alpha = -\frac{1}{L} \times \ln\left(\frac{1 - \kappa}{D^2}\right) \tag{2.16}$$

This relationship given by Eq. (2.16) is also referred to as critical coupling, where the maximum on-off ratio $\frac{I_t}{I_i}(2 m\pi) = 0$ can be obtained by varying the coupling coefficient (κ) or the intensity attenuation coefficient (α).

2.3 Resonance Bandwidth

Resonance bandwidth determines how fast optical data can be processed by a ring resonator. The resonator bandwidth is given by the full-width at half-maximum (FWHM or 3 dB bandwidth) $\delta\phi$ [$I_t/I_i(\varphi) = 0.5$] and the finesse F of the resonator is given by:

$$\delta\phi = \frac{2(1 - xy)}{\sqrt{xy}} \tag{2.17}$$

To understand how the bandwidth of the resonator is affected by the coupling coefficient κ, we will consider critically coupled ring resonator [23, 24]. In such a case,

$$\delta\phi = \frac{2k}{\sqrt{1 - k}} \tag{2.18}$$

Therefore, the lower coupling coefficient, the smaller resonance bandwidth is obtained.

2.4 Finesse

The finesse of the resonator is defined as a ratio of the free spectral range and the full width at half maximum of the resonance. For the Fig. 2.2 using FSR (frequency spacing between two resonance) in terms of the is equal to 2π and thus the finesse is given by [25–27]

$$F = \frac{2\pi}{\delta\phi} = \frac{\pi\sqrt{xy}}{1 - xy} \tag{2.19}$$

2.5 Free Spectral Range (FSR)

The frequency spacing between two resonance peaks is called the free spectral range which can be calculated. The phase constant which corresponds to $\phi = 2(m + 1)\pi$ is defined as $\mathbb{Q}$. The phase constant corresponds to $\phi = 2(m + 1)\pi$ is defined as $\mathbb{Q} + \Delta\mathbb{Q}$. The frequency shift Δf and the wavelength shift $\Delta\lambda$ are related to the variation of the phase constant $\Delta\mathbb{Q}$ as $\Delta f = (c/2\pi) \cdot \Delta\mathbb{Q}$ and $\Delta\lambda = -(\lambda^2/2\pi) \cdot \Delta\mathbb{Q}$. The resonance spacing in terms of the frequency f and the wavelength λ are given by [28, 29]

$$\Delta f = \frac{c}{n_{gr} \cdot L} \tag{2.20}$$

and

$$\Delta\lambda = \left| -\frac{\lambda^2}{n_{gr} \cdot L} \right| \tag{2.21}$$

where n_{gr} is the group refractive index, which is defined as;

$$n_{gr} = n_{eff} - \lambda \frac{dn_{eff}}{d\lambda} | \tag{2.22}$$

2.6 Quality Factor

Another value for characterization of ring resonator is the Q factor, The Q factor of the resonator is a measure of the sharpness of the resonance. In analogy with electrical circuit, the quality factor of an optical waveguide due it stored energy and the power lost per optical cycle. The Q factor is defined as [30–32]

$$Q = \omega \frac{stored\ energy}{Power\ Loss} \tag{2.23}$$

where ω is the frequency of the light coupled to the resonator. The Q factor of the resonator can be calculated from.

$$Q = \frac{f_0}{\delta f} = \frac{\lambda_0}{\delta\lambda} \tag{2.24}$$

The Q factor is the ratio of the absolute frequency f_0 or absolute wavelength λ_0 to the 3 dB bandwidth (δf or $\delta\lambda$). The shape and the bandwidth of the fiber response is determined by Q factor. The finesse and the Q factor are both important when one is interested in both the FSR (Δf or $\Delta \lambda$) and the 3 dB bandwidth (δf or $\delta \lambda$). They are related by

$$\frac{Q}{F} = \frac{f_0}{\Delta f} = \frac{\lambda_0}{\Delta \lambda} \qquad (2.25)$$

2.7 Chaotic Soliton Signal Generator

SMRR consists of a single coupler and a micoring resonator. The nonlinearity of the fiber ring is of the Kerr-type wherein the nonlinear refractive index is given by [33, 34]

$$n = n_0 + n_2 I = n_0 + \left(\frac{n_2}{A_{eff}}\right) P, \qquad (2.26)$$

where n_0 and n_2 are the linear and nonlinear refractive indices, respectively. I and P are the optical intensity and optical field power, respectively. The effective mode core area of the fiber is A_{eff}. Schematic of SMRR is illustrated in Fig. 2.4.

Input light of monochromatic Gaussian laser beam is introduced into the system. The fiber has a nonlinear refractive index of n_2 and a linear absorption coefficient of α. The intensity coupling coefficient of the fiber coupler is κ, where γ is a coupling loss of the field amplitude. The fiber ring has resonant condition for the specific wavelength in linear case. Here the fiber coupler is considered as a point device and is reciprocal. The relation between the electric fields E_1 and E_2, can be expressed using the nonlinear form as [35, 36]:

$$E_2 = E_1 x \exp\{-j(\phi_0 + \phi_{NL})\}, \qquad (2.27)$$

where $\phi_0 = kLn_0$ and $\phi_{NL} = kLn_2|E_1|^2$ are expressed as linear and nonlinear phase shift, $k = 2\pi/\lambda$ is a wave number and L is the circumference of the ring resonator. $x = \exp(-\alpha L/2)$ represents a round trip loss for the input pulse propagating inside

Fig. 2.4 Nonlinear silicon microring resonator (SMRR)

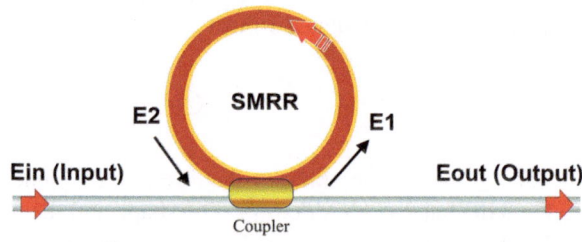

the SMRR. Mathematically, the subsequence Equations of the round trip within the system is given by [37, 38]

$$E_{n+1} = j\sqrt{(1-\gamma)\kappa}E_{in} + \sqrt{(1-\gamma)(1-\kappa)}xE_n \exp(-j(\phi_0 + \phi_{NL})), \quad (2.28)$$

where the subscript "n" denotes the number of round trip inside the SMRR. This equation has to be satisfied with boundary conditions appropriate to SMRR. In order to definiteness, we consider a SMRR connected to a single coupler that extracts light from the ring into the output waveguides, as schematically shown in Fig. 2.4. Here, general case with external field, injected into the SMRR has been studied. To have more simplicity, we consider that the coupler device is ideal, where it simply splits the input fields without internal losses at the operating wavelength. We also ignore reflectivities at the coupler–waveguide interface, which is usually a good approximation due to the same structure of the output waveguides and the coupler. Regards to steady situation, the output field can be expressed as [39–41]:

$$E_{out} = \sqrt{1-\gamma} \cdot E_{in} \left[\sqrt{1-\kappa} - \frac{\sqrt{1-\gamma}\,\kappa x \, \exp\left(-j(\phi_0 + \phi_{NL})\right)}{1 - \sqrt{(1-\gamma)(1-\kappa)}\,x \, \exp\left(-j(\phi_0 + \phi_{NL})\right)} \right].$$
$$(2.29)$$

Thus the output power of the light field is given by

$$P_{out} = \left| E_{out} \cdot E_{out}^* \right| \qquad (2.30)$$

Equation (2.29) is mathematical relation used for characterizing of nonlinear effects of the ring resonator system. Therefore, nonlinear refractive index, coupling coefficient, and the radius of the SMRR system are variable parameters that cause the nonlinear behavior to be seen in different roundtrip and input power.

MRR's input optical power can be in the form of an optical soliton pulse or a laser Gaussian beam expressed by Eqs. 2.31 and 2.32. The advantage of using optical soliton is that, it can be used to generate chaotic filter characteristics while propagating within the MRRs [42]. When the input pulse is brought into the MRRs system shown in Fig. 2.6.

$$E_{in} = A \tan h \left[\frac{T}{T_0} \right] \exp \left[\left(\frac{z}{2L_D} \right) - i\omega_0 t \right] \qquad (2.31)$$

$$E_{in}(t) = E_0 \exp^{j\phi_0(t)} \qquad (2.32)$$

A and z are optical field amplitude and propagation distance, respectively. T is the time of soliton pulse propagation, where $L_D = T_0^2 / |\beta_2|$ is the dispersion length of the soliton pulse and, β_2 is the propagation constant. Therefore, a soliton pulse describes a pulse, which keeps its temporal or spatial width invariance while it propagates inside the MRRs system. Soliton peak intensity is expressed by

$(|\beta_2/\Gamma T_0^2|)$, where $\Gamma = n_2 \times k_0$, is the length scale over which dispersive or nonlinear effects causes the beam to be wider or narrower. For a temporal soliton pulse in the micro ring device, a balance should be achieved between the dispersion length (L_D) and the nonlinear length $(L_{NL} = (1/\gamma\varphi_{NL})$, where γ and φ_{NL} are a coupling loss of the field amplitude and nonlinear phase shift. Here $\phi_0 = kLn_0$ is linear phase shifts. The refractive index (n) is given by [43, 44]:

$$n = n_0 + n_2 I = n_0 + \left(\frac{n_2}{A_{eff}}\right) P, \tag{2.33}$$

In which n_0 and n_2 are the linear and nonlinear refractive indices, respectively. I is the optical intensity and P is the optical power. For the micro ring and nano-ring resonators, the effective mode core area ranges from 0.50 to 0.1 μm^2. MRRs proposed system is consisting of series of ring resonators. This system can be connected to a rotatable polarizer and a beam splitter, which is used to generate distinct optical soliton pulses and quantum codes with greeter free spectrum range (FSR), shown by Fig. 2.5.

When the input pulse introduced into the MRRs system as shown in Fig. 2.5, the resonant output is built up in the series of MRRs, where Eq. (2.34) can express the normalized output of the light field [45, 46].

$$\left|\frac{E_{out}(t)}{E_{in}(t)}\right|^2 = (1-\gamma)\left[1 - \frac{(1-(1-\gamma)\,x^2)\kappa}{(1 - x\sqrt{1-\gamma}\sqrt{1-\kappa})^2 + 4\,x\sqrt{1-\gamma}\sqrt{1-\kappa}\sin^2(\frac{\phi_o + \phi_{NL}}{2})}\right] \tag{2.34}$$

Equation (2.34) points that a ring resonator in this exacting case is comparable to a Fabry-Perot cavity which is consisting of an input and output mirror with a field reflectivity, $(1 - \kappa)$, and a fully reflecting mirror. Equation (2.34) shows the output power from each ring resonator, which is realized as input power for the next ring resonator. Simulation results of the mathematical equations provide applicable

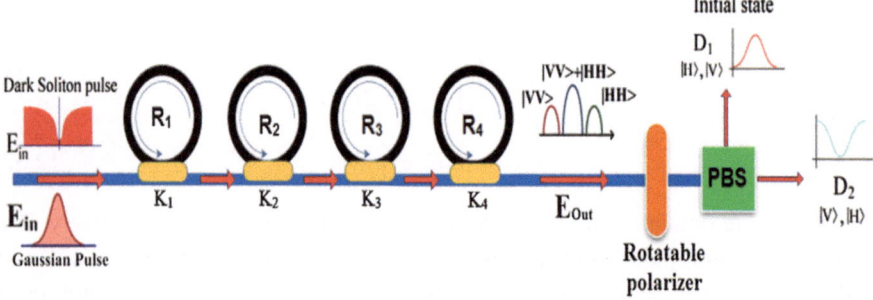

Fig. 2.5 System of a discrete pulses generation, where *PBS* polarizing Beam splitter, *Ds* detectors, *Rs* ring radii and κ_s coupling coefficients

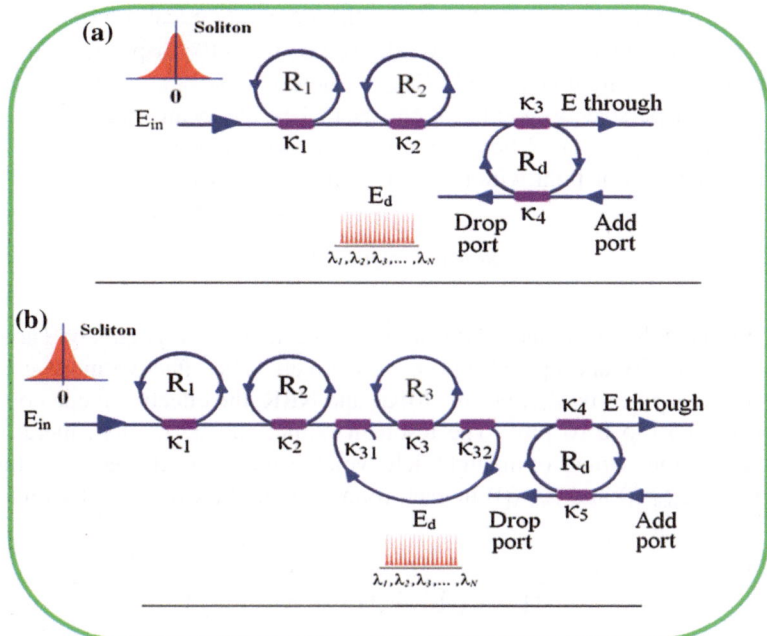

Fig. 2.6 Systems of multi optical soliton pulse generation **a** multi soliton generation, **b** multi soliton trapping and storage, R_s ring radii, κ_s coupling coefficients, κ_{31} and κ_{32} are coupling losses

optical soliton pulses to generate quantum codes, used for the secured networks communication.

When an optical soliton pulse input into the nonlinear MRR, the large optical bandwidth of the output signals can be generated, where the nonlinear behavior of self-phase modulation (SPM) keeps the large output power. Chaotic signals cancelation can be done using an optical add/drop filter system. The schematic of the two proposed systems are shown in Fig. 2.6.

The bright soliton pulse is introduced into the proposed system. The input optical field (E_{in}) of the optical bright soliton can be expressed as,

$$E_{in} = A \sec h \left[\frac{T}{T_0}\right] \exp\left[\left(\frac{z}{2L_D}\right) - i\omega_0 t\right] \qquad (2.35)$$

A and z are the optical field amplitude and propagation distance, respectively. T is a soliton pulse propagation time in a frame moving at the group velocity, $T = t - \beta_1 \times z$, where β_1 and β_2 are the coefficients of the linear and second order terms of Taylor expansion of the propagation constant. $L_D = T_0^2/|\beta_2|$ is the dispersion length of the soliton pulse [47, 48]. The frequency shift of the soliton is ω_0. This solution describes a pulse that keeps its temporal width invariance as it propagates, and thus is called a temporal soliton. When a soliton peak intensity

$(|\beta_2/\Gamma T_0^2|)$ is given, then T_o is known. For the temporal optical soliton pulse in the microring device, a balance should be achieved between the dispersion length (L_D) and the nonlinear length ($L_{NL} = (1/\ \Gamma\phi_{NL})$, where $\Gamma = n_2 \times k_0$, is the length scale over which dispersive or nonlinear effects makes the beam becomes wider or narrower, hence $L_D = L_{NL}$. When light propagates within the nonlinear medium, the refractive index (n) of light within the medium is given by [49]

$$n = n_0 + n_2 I = n_0 + \left(\frac{n_2}{A_{eff}}\right)P, \qquad (2.36)$$

n_0 and n_2 are the linear and nonlinear refractive indexes, respectively. I and P are the optical intensity and optical power, respectively. The effective mode core area of the device is given by A_{eff}. For the MRR and NRR, the effective mode core areas range from 0.50 to 0.10 μm^2. The resonant output can be formed; therefore the normalized output signals of the light field which is the ratio between the output and input fields (E_{out} (t) and E_{in} (t)) in each roundtrip can be expressed by [50, 51]

$$\left|\frac{E_{out}(t)}{E_{in}(t)}\right|^2 = (1-\gamma)\left[1 - \frac{(1-(1-\gamma)x^2)\kappa}{(1 - x\sqrt{1-\gamma}\sqrt{1-\kappa})^2 + 4x\sqrt{1-\gamma}\sqrt{1-\kappa}\sin^2(\frac{\phi}{2})}\right]$$
$$(2.37)$$

In this investigation, the iterative method is introduced to obtain the results as shown in Eq. (2.37), similarly, when the output field is connected and input into the next ring resonators. In order to retrieve the signals from the chaotic noise, we propose to use the add/drop device with the appropriate parameters. The optical outputs of a ring resonator add/drop filter are given by.

$$\left|\frac{E_t}{E_{in}}\right|^2 = \frac{(1-\kappa_1) - 2\sqrt{1-\kappa_1}\cdot\sqrt{1-\kappa_2}e^{-\frac{\alpha}{2}L}\cos(k_nL) + (1-\kappa_2)e^{-\alpha L}}{1 + (1-\kappa_1)(1-\kappa_2)e^{-\alpha L} - 2\sqrt{1-\kappa_1}\cdot\sqrt{1-\kappa_2}e^{-\frac{\alpha}{2}L}\cos(k_nL)} \qquad (2.38)$$

and

$$\left|\frac{E_d}{E_{in}}\right|^2 = \frac{\kappa_1\kappa_2 e^{-\frac{\alpha}{2}L}}{1 + (1-\kappa_1)(1-\kappa_2)e^{-\alpha L} - 2\sqrt{1-\kappa_1}\cdot\sqrt{1-\kappa_2}e^{-\frac{\alpha}{2}L}\cos(k_nL)} \qquad (2.39)$$

E_t and E_d represent the optical fields of the through port and drop ports, respectively. $\beta = kn_{eff}$ is the propagation constant, n_{eff} is the effective refractive index of the waveguide, and the circumference of the ring is $L = 2\pi R$, with R as the radius of the ring. New parameters are introduced for simplification with $\phi = \beta L$ as the phase constant. By using the specific parameters of the add/drop device, the chaotic noise cancellation can be obtained and the required signals can be retrieved by the specific users. κ_1 and κ_2 are the coupling coefficients of the add/drop filters, $k_n = 2\pi/\lambda$ is the wave propagation number in a vacuum, and the waveguide (ring

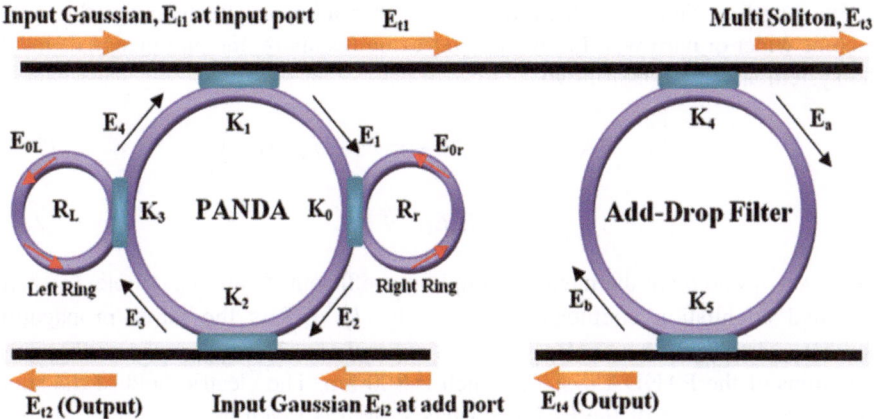

Fig. 2.7 Schematic diagram of a PANDA ring resonator connected to an add/drop filter system

resonator) loss is $\alpha = 0.5$ dBmm^{-1}. The fractional coupler intensity loss is $\gamma = 0.1$. In the case of the add/drop device, the nonlinear refractive index is neglected. High capacity of optical pulses can be obtained when the full width at half maximum (FWHM) of these pulses are very small, where the intensity build up can be performed inside the micro or nanoring system. The proposed system consists of a PANDA ring resonator connected to an add/drop filter system, shown in Fig. 2.7.

The laser Gaussian pulse input propagates inside the ring resonators system which is introduced by the nonlinear Kerr effect. The Kerr effect causes the refractive index (n) of the medium to vary as shown by

$$n = n_0 + n_2 I = n_0 + \frac{n_2}{A_{eff}} P, \qquad (2.40)$$

where n_0 and n_2 are the linear and nonlinear refractive indexes, respectively. I and P are the optical intensity and the power, respectively. A_{eff} is the effective mode core area of the device. For an add/drop optical filter design, the effective mode core areas range from 0.50 to 0.10 μm^2. The parameters were obtained by using practical parameters of used material (InGaAsP/InP) [52, 53]. Input optical fields of Gaussian pulses at the input and add ports of the system are given in Eq. (2.41).

$$E_{i1}(t) = E_{i2}(t) = E_{i0} \exp\left[\left(\frac{z}{2L_D}\right) - i\omega_0 t\right], \qquad (2.41)$$

E_i and z are the optical field amplitude and propagation distance respectively. L_D is the dispersion length of the soliton pulse where t is the soliton phase shift time, and where the carrier frequency of the signal is ω_0. Soliton pulses propagate within the microring device when the balance between the dispersion length (L_D) and the nonlinear length ($L_{NL} = 1/\Gamma\phi_{NL}$) is achieved. Therefore $L_D = L_{NL}$, where $\Gamma = n_2 \times k_0$, is the length scale over which dispersive or nonlinear effects make the beam become wider or narrower. For the PANDA ring resonator, the output signals inside the system are given as follows:

$$E_1 = \sqrt{1 - \gamma_1}\left(\sqrt{1 - \kappa_1}E_4 + j\sqrt{\kappa_1}E_{i1}\right) \tag{2.42}$$

$$E_2 = E_0E_1e^{-\frac{\alpha L}{22}-jk_n\frac{L}{2}}, \tag{2.43}$$

where κ_1, γ_1 and α are the intensity coupling coefficient, fractional coupler intensity loss and attenuation coefficient respectively. $k_n = \frac{2\pi}{\lambda}$ is the wave propagation number, λ is the input wavelength light field and $L = 2\pi R_{PANDA}$ where, R_{PANDA} is the radius of the PANDA system, which is 300 nm. The electric field of the small ring at the right side of the PANDA rings system is given as [54, 55]:

$$E_0 = E_1 \frac{\sqrt{(1 - \gamma)(1 - \kappa_0)} - (1 - \gamma)e^{-\frac{\alpha}{2}L_1 - jk_nL_1}}{1 - \sqrt{1 - \gamma}\sqrt{1 - \kappa_0}e^{-\frac{\alpha}{2}L_1 - jk_nL_1}}, \tag{2.44}$$

where $L_1 = 2\pi R_r$ and R_r is the radius of right ring. Light fields of the left side of PANDA ring resonator can be expressed as:

$$E_3 = \sqrt{1 - \gamma_2}\left[\sqrt{1 - \kappa_2}E_2 + j\sqrt{\kappa_2}E_{i2}\right], \tag{2.45}$$

$$E_4 = E_{0L}E_3e^{-\frac{\alpha L}{22}-jk_n\frac{L}{2}}, \tag{2.46}$$

where,

$$E_{0L} = E_3 \frac{\sqrt{(1 - \gamma)(1 - \kappa_3)} - (1 - \gamma)e^{-\frac{\alpha}{2}L_2 - jk_nL_2}}{1 - \sqrt{1 - \gamma}\sqrt{1 - \kappa_3}e^{-\frac{\alpha}{2}L_2 - jk_nL_2}}, \tag{2.47}$$

Here, $L_2 = 2\pi R_L$ and R_L is the left ring radius. In order to simplify these equations, the parameters of x_1, x_2, y_1 and y_2 are defined as: $x_1 = \sqrt{(1 - \gamma_1)}$, $x_2 = \sqrt{(1 - \gamma_2)}$, $y_1 = \sqrt{(1 - \kappa_1)}$, and $y_2 = \sqrt{(1 - \kappa_2)}$. Therefore, the output powers from through and drop ports of the PANDA ring resonator can be expressed as [56–58]

$$E_{t1} = AE_{i1} - BE_{i2}e^{-\frac{\alpha L}{22}-jk_n\frac{L}{2}}\left[\frac{CE_{i1}G^2 + DE_{i2}G^3}{1 - FG^2}\right], \tag{2.48}$$

$$E_{t2} = x_2y_2E_{i2}\left[\frac{A\sqrt{\kappa_1\kappa_2}E_0E_{i1}G + \frac{D}{x_1\sqrt{\kappa_1E_{0L}}}E_{i2}G^2}{1 - FG^2}\right], \tag{2.49}$$

where, $\quad A = x_1 x_2, \qquad B = x_1 x_2 y_2 \sqrt{\kappa_1} E_{0L}, \qquad C = x_1^2 x_2 \kappa_1 \sqrt{\kappa_2} E_0 \dot{E}_{0L},$

$D = (x_1 x_2)^2 y_1 y_2 \sqrt{\kappa_1 \kappa_2} E_0 E_{0L}^2, \ G = \left(e^{-\frac{\alpha L}{22} - jk_n \frac{L}{2}} \right)$ and $F = x_1 x_2 y_1 y_2 E_0 E_{0L}.$

E_{t1}, output from the PANDA system can be input into the add/drop filter system which is made of a ring resonator coupled to two fiber waveguides with proper parameters.

Output powers from the add/drop filter system are given by Eqs. (2.50) and (2.51), where E_{t3} and E_{t4} are the electric field outputs of the through and drop ports of the system respectively.

$$\frac{I_{t3}}{I_{t1}} = \left| \frac{E_{t3}}{E_{t1}} \right|^2 = \frac{1 - \kappa_4 - 2\sqrt{1 - \kappa_4}\sqrt{1 - \kappa_5} e^{\frac{-\alpha}{2} L_{ad}} \cos(k_n L_{ad}) + (1 - \kappa_5) e^{-\alpha L_{ad}}}{1 + (1 - \kappa_4)(1 - \kappa_5) e^{-\alpha L_{ad}} - 2\sqrt{1 - \kappa_4}\sqrt{1 - \kappa_5} e^{\frac{-\alpha}{2} L_{ad}} \cos(k_n L_{ad})}$$

(2.50)

$$\frac{I_{t4}}{I_{t1}} = \left| \frac{E_{t4}}{E_{t1}} \right|^2 = \frac{\kappa_4 \cdot \kappa_5 e^{\frac{-\alpha}{2} L_{ad}}}{1 + (1 - \kappa_4)(1 - \kappa_5) e^{-\alpha L_{ad}} - 2\sqrt{1 - \kappa_4}\sqrt{1 - \kappa_5} e^{\frac{-\alpha}{2} L_{ad}} \cos(k_n L_{ad})}$$

(2.51)

where, κ_4 and κ_5 are the coupling coefficients of the add/drop filter system, $L_{ad} = 2\pi R_{ad}$ and R_{ad} is the radius of the add/drop system.

References

1. V.N. Serkin, A. Hasegawa, Novel soliton solutions of the nonlinear Schrödinger equation model. Phys. Rev. Lett. **85**(21), 4502 (2000)
2. P. Wai, C. Menyuk, H. Chen, Y. Lee, Soliton at the zero-group-dispersion wavelength of a single-model fiber. Opt. Lett. **12**(8), 628–630 (1987)
3. P. Beaud, W. Hodel, B. Zysset, H. Weber, Ultrashort pulse propagation, pulse breakup, and fundamental soliton formation in a single-mode optical fiber. IEEE J. Quantum Electron. **23** (11), 1938–1946 (1987)
4. J. Yang, Z.H. Musslimani, Fundamental and vortex solitons in a two-dimensional optical lattice. Opt. Lett. **28**(21), 2094–2096 (2003)
5. D. Mihalache, D. Mazilu, L.-C. Crasovan, I. Towers, A. Buryak, B. Malomed, L. Torner, J. Torres, F. Lederer, Stable spinning optical solitons in three dimensions. Phys. Rev. Lett. **88** (7), 073902 (2002)
6. A. Afroozeh, I.S. Amiri, K. Chaudhary, J. Ali, P.P. Yupapin, Analysis of optical ring resonator. in *Advances in Laser and Optics Research* (Nova Science, New York, 2014)
7. I.S. Amiri, A. Afroozeh, I.N. Nawi, M.A. Jalil, A. Mohamad, J. Ali, P.P. Yupapin, Dark Soliton Array for communication security. Proc. Eng. **8**, 417–422 (2011)
8. A. Afroozeh, I.S. Amiri, A. Zeinalinezhad, *Micro Ring Resonators and Applications* (LAP LAMBERT Academic Publishing, Saarbrücken, 2014)
9. I.S. Amiri, S.E. Alavi, F.J. Rahim, S.M. Idrus, Analytical treatment of the ring resonator passive systems and bandwidth characterization using directional coupling coefficients. J. Comput. Theor. Nanosci. (2014)

10. I.S. Amiri, S.E. Alavi, M. Bahadoran, A. Afroozeh, S.M. Idrus, Nanometer bandwidth soliton generation and experimental transmission within nonlinear fiber optics using an add-drop filter system. J. Comput. Theor. Nanosci. (2014)
11. I.S. Amiri, S.E. Alavi, S.M. Idrus', Solitonic pulse generation and characterization by integrated ring resonators, in *5th International Conference on Photonics 2014 (ICP2014)*, Kuala Lumpur (2014)
12. I.S. Amiri, S.E. Alavi, S.M. Idrus, M. Kouhnavard, *Microring Resonator for Secured Optical Communication*. Amazon, Seattle (2014)
13. A. Nikoukar, I.S. Amiri, S.E. Alavi, A. Shahidinejad, T. Anwar, A.S.M. Supa'at, S.M. Idrus, L.Y. Teng, Theoretical and simulation analysis of the add/drop filter ring resonator based on the Z-transform method theory, in *The 2014 Third ICT International Student Project Conference (ICT-ISPC2014)*, Thailand (2014)
14. I.S. Amiri, K. Raman, A. Afroozeh, M.A. Jalil, I.N. Nawi, J. Ali, P.P. Yupapin, Generation of DSA for security application. Proc. Eng. **8**, 360–365 (2011)
15. I.S. Amiri, S.E. Alavi, S.M. Idrus, Results of digital soliton pulse generation and transmission using microring resonators, in *Soliton Coding for Secured Optical Communication Link* (Springer, New York, 2015), pp. 41–56.
16. I.S. Amiri, R. Ahsan, A. Shahidinejad, J. Ali, P.P. Yupapin, Characterisation of bifurcation and chaos in silicon microring resonator. IET Commun. **6**(16), 2671–2675 (2012)
17. I.S. Amiri, M. Ebrahimi, A.H. Yazdavar, S. Gorbani, S.E. Alavi, Sevia M. Idrus, and J. Ali, Transmission of data with orthogonal frequency division multiplexing technique for communication networks using GHz frequency band soliton carrier. IET Commun. **8**(8), 1364–1373 (2014)
18. A. Jezierski, P. Laybourn, Integrated semiconductor ring lasers. IEE Proc. J. Optoelectron. 17–24 (1988)
19. I.S. Amiri, S.E. Alavi, S.M. Idrus, *Soliton Coding for Secured Optical Communication Link* (Springer, New York, 2014)
20. I.S. Amiri, FWHM measurement of localized optical soliton, in *The International Conference for Nano materials Synthesis and Characterization*, Malaysia (2011)
21. I.S. Amiri, S.E. Alavi, A. Shahidinejad, A. Nikoukar, T. Anwar, A.S.M. Supa'at, S.M. Idrus, N.K. Yen, Characterization of ultra-short soliton generation using MRRs, in *The 2014 Third ICT International Student Project Conference (ICT-ISPC2014)*, Thailand (2014)
22. I.S. Amiri, S.E. Alavi, S.M. Idrus, A.S.M. Supa'at, J. Ali, P.P. Yupapin, W-Band OFDM transmission for radio-over-fiber link using solitonic millimeter wave generated by MRR. IEEE J. Quantum Electron. **50**(8), 622–628 (2014)
23. I.S. Amiri. A. Afroozeh, *Ring Resonator Systems to Perform Optical Communication Enhancement Using Soliton* (Springer, New York, 2014)
24. I.S. Amiri, A. Afroozeh, Integrated Ring Resonator Systems, in *Ring Resonator Systems to Perform Optical Communication Enhancement Using Soliton* (Springer, New York, 2015), pp. 37–47
25. I.S. Amiri, A. Afroozeh, M. Bahadoran, Simulation and analysis of multisoliton generation using a PANDA ring resonator system. Chin. Phys. Lett. **28**(10), 104205 (2011)
26. I.S. Amiri, J. Ali, P.P. Yupapin, Enhancement of FSR and finesse using add/drop filter and PANDA ring resonator systems. Int. J. Mod. Phys. B **26**(04), 1250034 (2012)
27. I.S. Amiri, A. Afroozeh, Mathematics of soliton transmission in optical fiber, in *Ring Resonator Systems to Perform Optical Communication Enhancement Using Soliton* (Springer, New York, 2015), pp. 9–35
28. S.E. Alavi, I.S. Amiri, S.M. Idrus, A.S.M. Supa'at, J. Ali, P. P. Yupapin, All optical OFDM generation for IEEE802.11a based on soliton carriers using microring resonators. IEEE Photonics J. **6**(1) (2014)
29. I.S. Amiri, J. Ali, Generating highly dark-bright solitons by Gaussian beam propagation in a PANDA ring resonator. J. Comput. Theor. Nanosci. **11**(4), 1092–1099 (2014)

30. I.S. Amiri, J. Ali, Data signal processing via a Manchester coding-decoding method using chaotic signals generated by a PANDA ring resonator. Chin. Opt. Lett. **11**(4), 041901(4) (2013)

31. S.E. Alavi, I.S. Amiri, S.M. Idrus, A.S.M. Supa'at, Generation and wired/wireless transmission of IEEE802.16 m signal using solitons generated by microring resonator. Opt. Quantum Electron. (2014)

32. I. Wolff, Microstrip bandpass filter using degenerate modes of a microstrip ring resonator. Electron. Lett. **8**(12), 302–303 (1972)

33. I.S. Amiri, S.E. Alavi, J. Ali, High capacity soliton transmission for indoor and outdoor communications using integrated ring resonators. Int. J. Commun. Syst. (2013)

34. I.S. Amiri, A. Afroozeh, Introduction of soliton generation, in *Ring Resonator Systems to Perform Optical Communication Enhancement Using Soliton* (Springer, New York, 2015), pp. 1–7

35. I.S. Amiri, M. Nikmaram, A. Shahidinejad, J. Ali, Generation of potential wells used for quantum codes transmission via a TDMA network communication system. Secur. Commun. Netw. **6**(11), 1301–1309 (2013)

36. I.S. Amiri, A. Afroozeh, Soliton generation based optical communication, in *Ring Resonator Systems to Perform Optical Communication Enhancement Using Soliton* (Springer, New York, 2015), pp. 49–68

37. I.S. Amiri, J. Ali, Optical quantum generation and transmission of 57–61 GHz frequency band using an optical fiber optics. J. Comput. Theor. Nanosci. **11**(10), 2130–2135 (2014)

38. I.S. Amiri, A. Nikoukar, S.E. Alavi, *Soliton and Radio over Fiber (RoF) Applications* (LAP LAMBERT Academic Publishing, Saarbrücken, 2014)

39. A. Afroozeh, I.S. Amiri, A. Zeinalinezhad, S.E. Pourmand, H. Ahmad, Comparison of control light using Kramers-Kronig method by three waveguides. J. Comput. Theor. Nanosci. (2015)

40. I.S. Amiri, S. Soltanmohammadi, A. Shahidinejad, J. Ali, Optical quantum transmitter with finesse of 30 at 800-nm central wavelength using microring resonators. Opt. Quantum Electron. **45**(10), 1095–1105 (2013)

41. I.S. Amiri, J. Ali, Picosecond Soliton pulse generation using a PANDA system for solar cells fabrication. J. Comput. Theor. Nanosci. **11**(3), 693–701 (2014)

42. I.S. Amiri, S.E. Alavi, S.M. Idrus, A. Afroozeh, J. Ali, *Soliton Generation by Ring Resonator for Optical Communication Application* (Nova Science Publishers, Hauppauge, 2014)

43. I.S. Amiri, J. Ali, Nano particle trapping by ultra-short tweezer and wells Using MRR interferometer system for spectroscopy application. Nanosci. Nanotechnol. Lett. **5**(8), 850–856 (2013)

44. I.S. Amiri, A. Afroozeh, Spatial and temporal soliton pulse generation by transmission of chaotic signals using fiber optic link, in *Advances in Laser and Optics Research*, vol. 11 (Nova Science Publisher, New York, 2014)

45. I.S. Amiri, J. Ali, Femtosecond optical quantum memory generation using optical bright soliton. J. Comput. Theor. Nanosci. **11**(6), 1480–1485 (2014)

46. S.E. Alavi, I.S. Amiri, S.M. Idrus, A.S.M. Supa'at, Optical amplification of tweezers and bright soliton using an interferometer ring resonator system. J. Comput. Theor. Nanosci. (2014)

47. S.E. Alavi, I.S. Amiri, A.S.M. Supa'at, S.M. Idrus, Indoor data transmission over ubiquitous infrastructure of powerline cables and LED lighting. J. Comput. Theor. Nanosci. (2014)

48. I.S. Amiri, M.H. Khanmirzaei, M. Kouhnavard, P.P. Yupapin, J. Ali, Quantum entanglement using multi dark soliton correlation for multivariable quantum router, in *Quantum Entanglement* ed. by A.M. Moran (Nova Science Publisher, New York, 2012), pp. 111–122

49. I.S. Amiri, Optical soliton trapping for quantum key generation, in *The International Conference for Nano materials Synthesis and Characterization*, Malaysia (2011)

50. P. Sanati, A. Afroozeh, I.S. Amiri, J. Ali, S.C Lee, Femtosecond pulse generation using microring resonators for eye nano surgery. Nanosci. Nanotechnol Lett. **6**(3), 221–226 (2014)

51. I.S. Amiri, B. Barati, P. Sanati, A. Hosseinnia, H.R. Mansouri Khosravi, S. Pourmehdi, A. Emami, J. Ali, Optical stretcher of biological cells using sub-nanometer optical tweezers

generated by an add/drop microring resonator system. Nanosci. Nanotechnol. Lett. **6**(2), 111–117 (2014)

52. I.S. Amiri, S.E. Alavi, S.M. Idrus, Introduction of fiber waveguide and soliton signals used to enhance the communication security, in *Soliton Coding for Secured Optical Communication Link* (Springer, New York, 2015), pp. 1–16

53. A. Zeinalinezhad, S.E. Pourmand, I.S. Amiri, A. Afroozeh, Stop light Generation using nano ring resonators for ROM. J. Comput. Theor. Nanosci. (2014)

54. I.S. Amiri, S.E. Alavi, S.M. Idrus, A. Nikoukar, J. Ali, IEEE 802.15.3c WPAN standard using millimeter optical soliton pulse generated by a PANDA ring resonator. IEEE Photonics J. **5**(5), 7901912 (2013)

55. I.S. Amiri, S.E. Alavi, S.M. Idrus, RF signal generation and wireless transmission using PANDA and add/drop systems. J. Comput. Theor. Nanosci. (2015)

56. I.S. Amiri, A. Nikoukar, J. Ali, GHz frequency band soliton generation using integrated ring resonator for WiMAX optical communication. Opt. Quantum Electron. (2013)

57. I.S. Amiri, P. Naraei, J. Ali, Review and theory of optical soliton generation used to improve the security and high capacity of MRR and NRR passive systems. J. Comput. Theor. Nanosci. **11**(9), 1875–1886 (2014)

58. I.S. Amiri, S.E. Alavi, S.M. Idrus, Theoretical background of microring resonator systems and soliton communication, in *Soliton Coding for Secured Optical Communication Link* (Springer, New York, 2015), pp. 17–39

Chapter 3
Solitonic Signals Generation and Transmission Using MRR

Abstract The advantage of the study is that nonlinear behavior of light inside ring resonators system can be controlled and managed by using appropriate parameters of the microring resonator (MRR) system. The effects of the coupling coefficients on the bifurcation and chaos behaviors, in terms of roundtrip and output powers are presented. Moreover, the radius of the ring is the key parameter which causes the nonlinear behaviors happen in lower roundtrips. Input optical powers can be in the form of an optical dark/bright soliton or a Gaussian beam. Optical soliton can be used to generate chaotic filter characteristics when propagating within the single microring resonator (SMRR). Therefore, chaotic signals are generated, whereas the required signals including specific wavelengths or frequencies can be used to perform the optical communication. The wide range of spread wavelength can be generated, using proposed MRRs system, where the wavelength multiplexing, dense wavelength division multiplexing (DWDM) can be perform via the optical wireless communication. The proposed MRR systems are suitable for the multi soliton pulses generation, which are available for high performance communication network. Increasing in communication capacity is provided by increasing of soliton pulses, which can be performed by generation of temporal and spatial optical soliton pulses. Generated multi optical soliton can be input into an optical receiver unit which is a quantum processing system used to generate high capacity packet of quantum information within the series of MRRs.

Keywords Microring resonator (MRR) · Bifurcation and chaos behaviors · Dark/bright soliton · Optical communication link · Communication capacity

3.1 Bifurcation and Chaos Behaviors Within the MRR

Considering the output resonant pulses, since the depth of this resonance dip decreases sharply when the round-trip losses increase (light needs to do many round-trips in the ring in order to exhibit a strong resonating behavior), it is necessary to keep these losses within a reasonable range or increase the input power.

© The Author(s) 2015
I. Sadegh Amiri and H. Ahmad, *Optical Soliton Communication Using Ultra-Short Pulses*, SpringerBriefs in Applied Sciences and Technology, DOI 10.1007/978-981-287-558-7_3

Therefore to control the roundtrips we need to control the loses which can be done by reducing the length of the ring or use the higher input power. In practical work the input power can be controlled by changing the current of the laser course. Lateral and vertical offsets allow control over the coupling coefficient. In this work the Iterative and Z-transformed methods are used in order to do the simulation and modelling using experimental parameters. Figure 3.1 shows the bifurcation and chaos behavior occurred for various nonlinear refractive indices of the system of single ring resonator. The parameters have been fixed to $\lambda_0 = 1.55$ μm, $n_0 = 3.37$, $A_{\text{eff}} = 0.30$ μm^2, $\alpha = 0.01$ dB km^{-1} and $\gamma = 0.1$. The length of the ring is $L = 80$ μm, where the coupling coefficients is $\kappa = 0.0225$ and the linear phase shift has been kept to zero. Total round trip of the input pulse inside the ring system is 20,000. An increasing of nonlinear refractive index from $n_2 = 2.2 \times 10^{-20}$ to $n_2 = 3.2 \times 10^{-20}$ m^2/W causes the optical nonlinear phenomena to be seen at the lower range of input power shown in Fig. 3.1.

The advantage of the study is that nonlinear behavior of light inside ring resonators system can be controlled and managed by using appropriate parameters of the ring system. Effects of the coupling coefficients on the bifurcation [1] and chaos behavior, in terms of roundtrip and output powers are shown in Fig. 3.2. Input Gaussian pulse with power of 2 W is introduced into the ring system where the radius of the ring is selected to $R = 15$ μm. Here, the coupling coefficient is a variable parameter that varies from $\kappa = 0.01$ to $\kappa = 0.1$.

Moreover, the radius of the ring is the key parameter which causes the bifurcation and chaos behavior to be happened in lower roundtrips [2–4]. These effects are shown in Fig. 3.3. Here the coupling coefficient has been fixed to $\kappa = 0.0225$, where the radius of the SMRR varies from 7 to 40 μm.

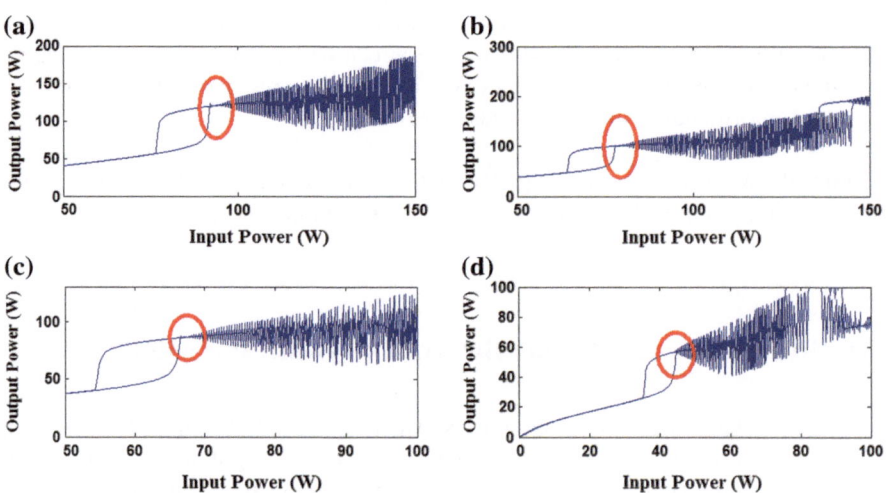

Fig. 3.1 Bifurcation and chaos behavior of light inside SMRR, where $\kappa = 0.0225$ and $L = 80$ μm for various nonlinear refractive indices: **a** $n_2 = 2.2 \times 10^{-20}$ m^2/W, **b** $n_2 = 2.4 \times 10^{-20}$ m^2/W, **c** $n_2 = 2.6 \times 10^{-20}$ m^2/W and **d** $n_2 = 3.2 \times 10^{-20}$ m^2/W

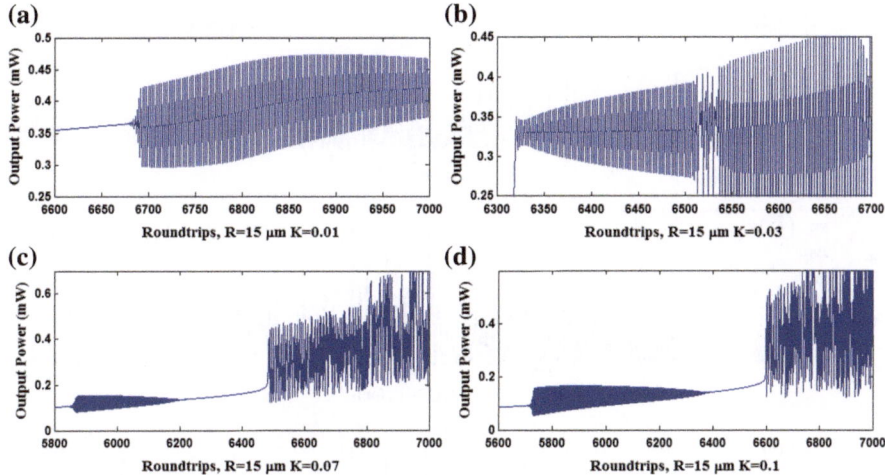

Fig. 3.2 Simulation results of bifurcation behavior generation within a SMRR respect to different value of couple coefficient (κ), where **a** $\kappa = 0.01$, **b** $\kappa = 0.03$, **c** $\kappa = 0.07$ and **d** $\kappa = 0.1$

In such a way, the bigger size of the ring resonator can be made which has lower cost compare to the smaller one [5–7]. Ring resonators can be fabricated and integrated in single system, easier when the ring of bigger size is used, applicable for fast switching optical devices. Therefore, chaotic signals are controlled and can be generated used to generate wireless signals which can be propagated inside computer network [8–10].

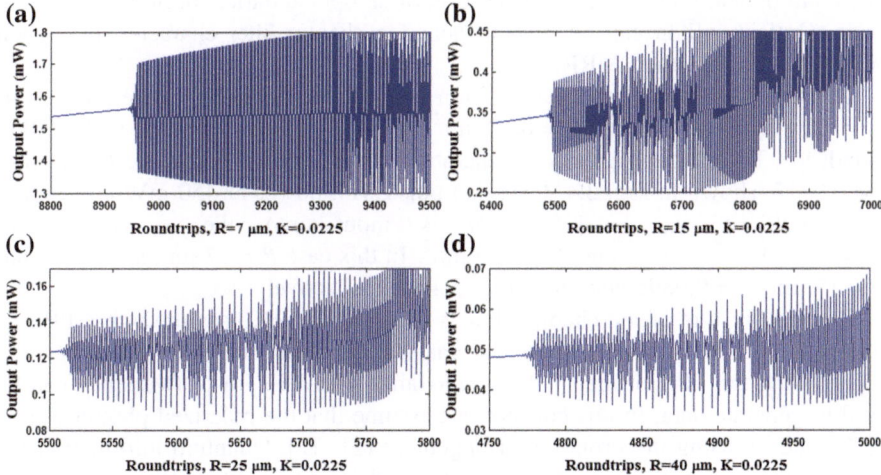

Fig. 3.3 Simulation results of bifurcation and chaos phenomena within a microring resonators respect to different values of ring radius (R)

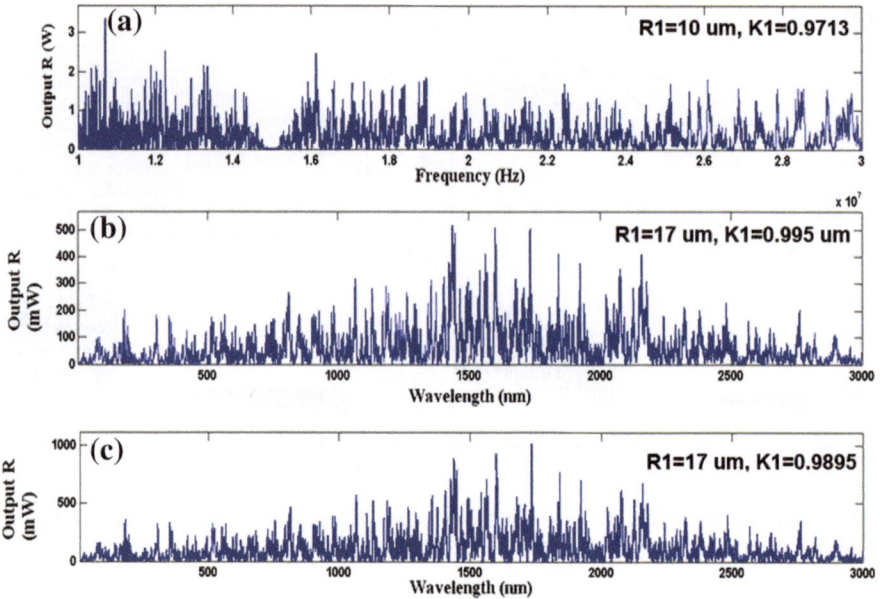

Fig. 3.4 Simulation results of chaotic signals within the SMRR, where **a** simulated frequency chaotic band, **b** and **c** spatial chaotic signals

3.2 Chaotic Signals Generated by MRR Used in Wireless Networks Systems

The input optical power can be in the form of an optical dark soliton or a Gaussian beam. Optical soliton can be used to generate chaotic filter characteristics when propagating within the SMRR.

The optical power of the dark soliton is fixed to 550 mW, where $n_0 = 3.34$, $n_2 = 2.2 \times 10^{-17}$ m^2 W^{-1}, $A_{\mathrm{eff}} = 0.50$ μm^2, $\alpha = 0.5$ dB mm^{-1}, $\gamma = 0.1$, with 20,000 roundtrips. The chaotic signals are generated within the ring (R), where $R = 10$ μm, $\kappa = 0.9713$, shown in Fig. 3.4. Gaussian pulse with power of 450 mW and central wavelength of $\lambda_0 = 1500$ nm is input into the system, where $n_2 = 2.2 \times 10^{-15}$ m^{2}W^{-1} and $A_{\mathrm{eff}} = 25$ μm^2. In this case $R = 17$ μm, $\kappa = 0.995$ and $R = 17$ μm, $\kappa = 0.9895$ shown in Fig. 3.4b, c respectively.

Obtained results of the chaotic signals from the SMRR pass through a PBS as shown in Fig. 3.5. In application, the variable quantum information can be generated using the PBS. It means that the localized wavelength or frequency can be used to generate data. In this concept, we assume that the polarized photon can be performed by using the proposed arrangement [11, 12]. Quantum information via chaotic signals can be connected into a network communication system shown in Fig. 3.5. Therefore, generated information can be transmitted to different users via a wireless networks transmitter system.

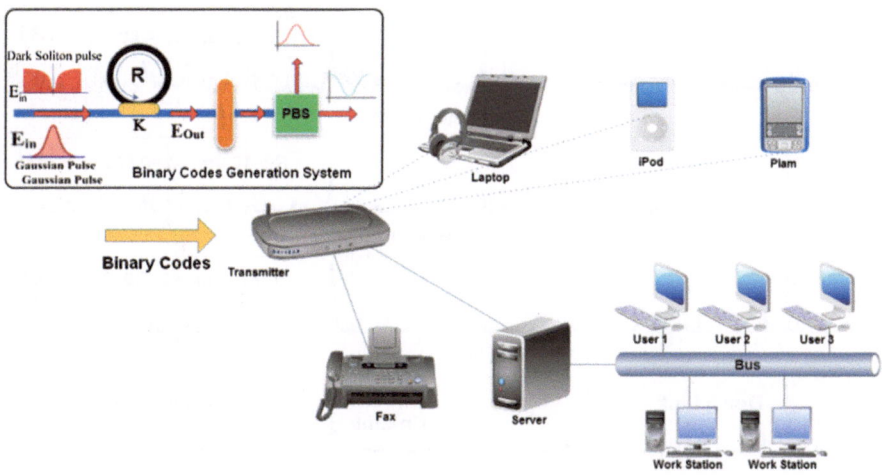

Fig. 3.5 Schematic of a computer wireless networks system, where the transmission of information in the form of binary codes can be implemented using SMRR

Therefore, chaotic signals are generated, whereas the required signals including specific wavelengths or frequencies can be used to perform the wireless communication network [13–15]. Thus, the information can be formed by the spatial soliton pulses generation. In order to increase the capacity of microring systems, more sharp optical pulses with smaller free spectrum range (FSR) are recommended. Furthermore, the applications such as quantum repeater, quantum entangled photon source are available, which can complete the concept of quantum optical communication networks [16, 17].

Multimode noisy signals can be generated from single mode of soliton pulse after circulating first MRR due to the nonlinear Kerr effect [18]. Large bandwidth of signals is generated, where the next ring resonators within the system form the chaotic filtering characteristics of the signals [19]. Using the semiconductor material, InGaAsp/InP, the specified frequency bands are obtained based on selected ring parameters [20, 21]. The soliton waveform with central frequency of $f_0 = 20$ MHz is input into the first MRR.

3.3 Up–Down-Link Frequency/Wavelength Soliton Band Generation

Using the series of MRRs, the optical power is fixed to 550 mW, where $n_0 = 3.34$, $n_2 = 2.2 \times 10^{-17}$ m^2 W^{-1}, $A_{\text{eff}} = 0.50$ μm^2, $\alpha = 0.5$ dB mm^{-1}, $\gamma = 0.1$, with 20,000 roundtrips. The chaotic signals are generated within the first ring (R_1), where the broad frequency band is observed in ring R_2. The filtering signals are seen in rings R_3 and R_4. Figure 3.6 shows the map of the simultaneous frequencies generation,

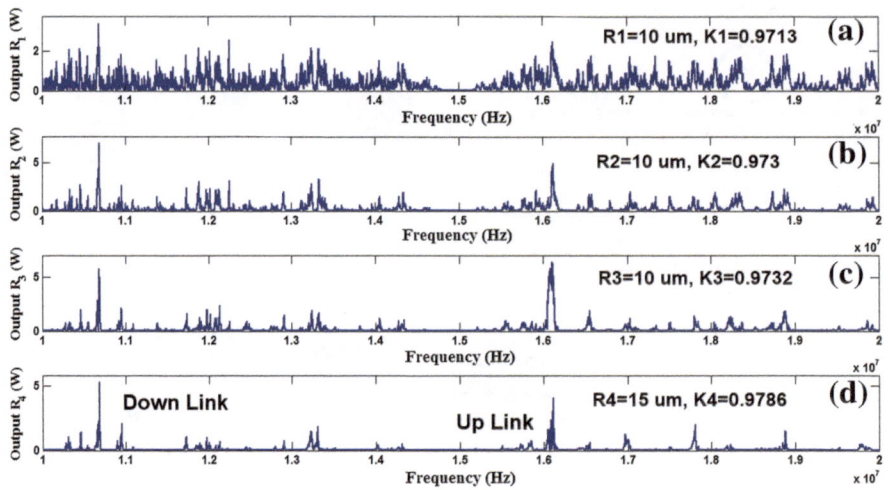

Fig. 3.6 Simultaneous of specific frequencies generation, **a** noisy chaotic signals, **b** frequency bands, **c** filtering signals, **d** discrete signals

for the ring parameters of $R_1 = 10$ μm, $\kappa_1 = 0.9713$, $R_2 = 10$ μm, $\kappa_2 = 0.973$, $R_3 = 10$ μm, $\kappa_3 = 0.9732$, $R_4 = 15$ μm, and $\kappa_4 = 0.9786$. In application, the upstream and downstream communication information can be linked via proposed system, for the up–down-link converters.

Discrete frequency bands from 10 to 30 MHz can be simultaneously generated, where the specified frequencies are filtered to form the required link converters. Figure 3.7 shows the frequency profile generated for a down-link converter.

Figure 3.8 shows the frequency generation for an up-link converter.

Wide range of spread wavelength can be generated, using proposed MRRs system, where the wavelength multiplexing, dense wavelength division multiplexing (DWDM) can be perform via optical wireless link [22, 23]. In principle, the specific wavelength mode is required in this technique; therefore, the chaotic signals are needed to be generated within the proposed [24, 25]. In this case, the input power is in the form of Gaussian beam and it is given by Eq. (2.32).

Gaussian pulse is input into the first MRR as shown in Fig. 3.9, where light modes can be generated with smaller spectral width than the input pulse. The optical filter characteristics is perform using appropriate ring parameters such as input power and, ring material, refractive index, radius and coupling constant, etc. Thus, Gaussian pulse with power of 450 mW and central wavelength of $\lambda_0 = 1500$ nm is input into the system, where $n_2 = 2.2 \times 10^{-15}$ m^2 W^{-1} and $A_{\text{eff}} = 25$ μm^2.

Figure 3.9 shows the trapping of wavelength pulses, obtained by using the MRRs system. Discrete wavelengths of $\lambda_D = 1448$ and $\lambda_U = 2169$ nm with powers of 592.6 and 394.6 mW are generated respectively. Here, the ring radii, R_1, R_2, R_3 and R_4 are 17, 13, 14 and 14 μm respectively. The coupling constants κ_1, κ_2, κ_3, κ_4 are selected to 0.995, 0.9831, 0.985 and 0.9826, respectively.

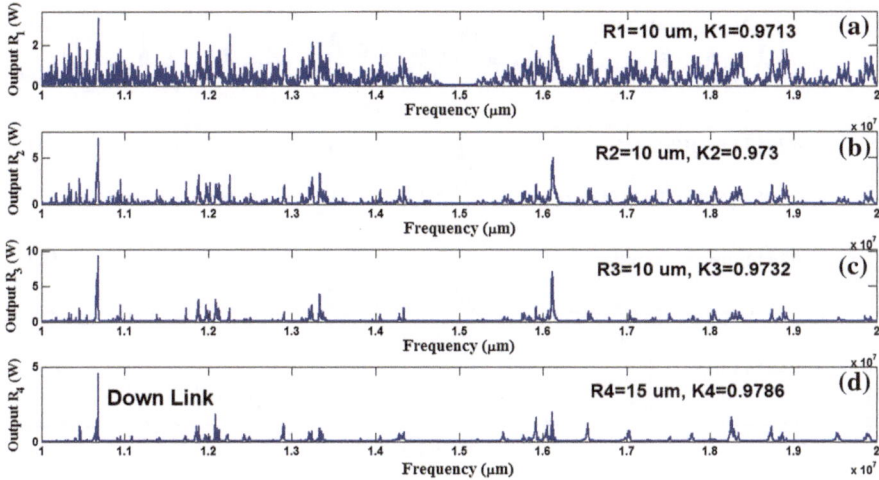

Fig. 3.7 Simulation of frequency band generation, **a** chaotic signals, **b** frequency bands, **c** filtering signals, **d** signal at 10.7 MHz

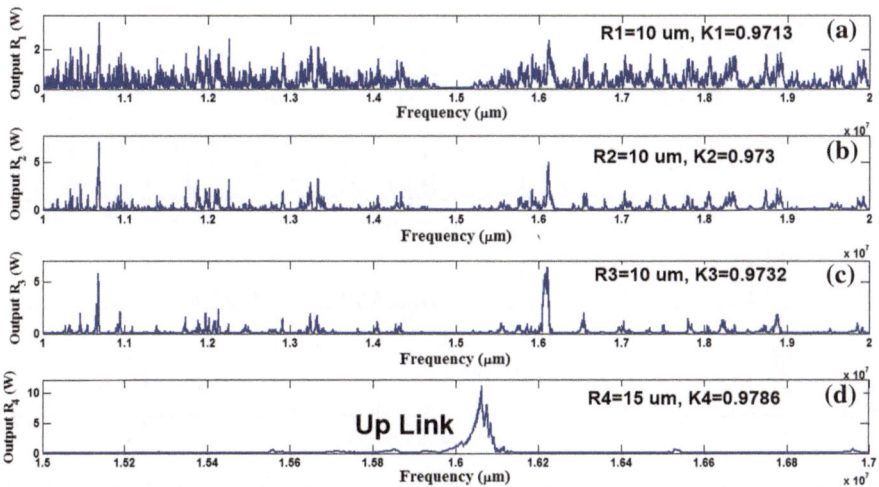

Fig. 3.8 Simulation of frequency band generation, **a** chaotic signals, **b** frequency bands, **c** filtering signals, **d** signal at 16 MHz

Figure 3.10 shows the filtering of chaotic signal, where the discrete wavelengths of $\lambda_D = 206.9$ and $\lambda_U = 2489$ nm with powers of 760.5 and 962.3 mW are generated respectively. Here, the coupling constants κ_1, κ_2, κ_3 and κ_4 of the rings have been selected to 0.9895, 0.9858, 0.9858 and 0.9713, respectively.

Obtained results of discrete wavelengths or frequencies from MRRs system can be passed through a PBS. In application, the quantum information can be generated

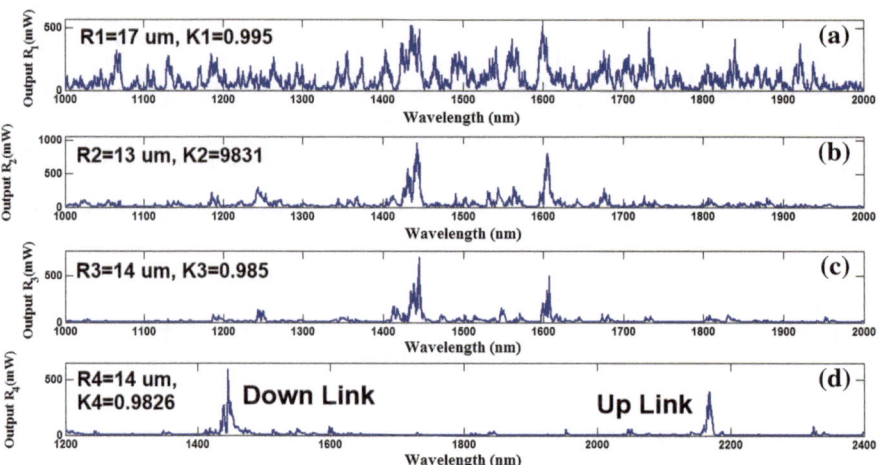

Fig. 3.9 Discrete wavelengths generation, where **a** chaotic signals from R_1, **b** filtering by R_2, **c** wavelength trapping by R_3, **d** localized wavelengths at $\lambda_D = 1448$ and $\lambda_U = 2169$ nm with powers of 592.6 and 394.6 mW respectively

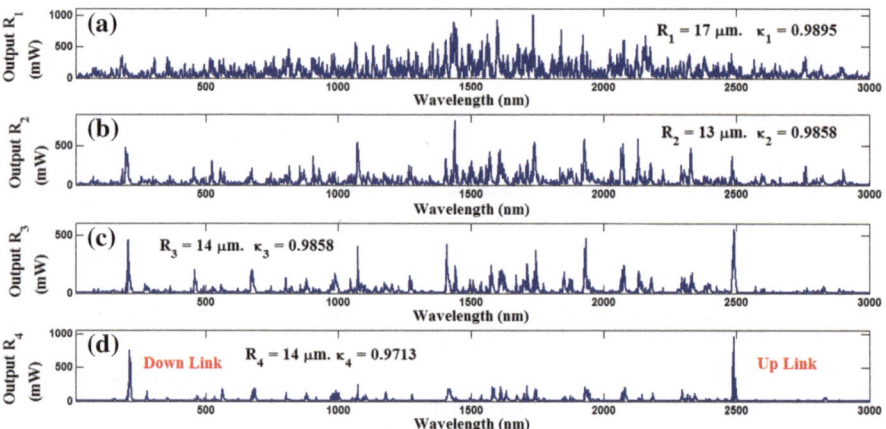

Fig. 3.10 Discrete wavelengths generation, where **a** chaotic signals from R_1, **b** filtering by R_2, **c** wavelength trapping by R_3, **d** localized wavelengths at $\lambda_D = 206.9$ nm and $\lambda_U = 2489$ nm with powers of 760.5 and 962.3 mW respectively

using the PBS [26, 27]. The used beam splitters reflect (and transmit) 50 % of the light that is incident, for all polarizations of the incident light. This interconnection can also be done with fiber couplers. Therefore, localized wavelength or frequency can be used to generate quantum information [28]. In this concept, we assume that the polarized photon can be performed by using the proposed arrangement.

Thus, localized optical soliton pulse are generated, whereas the required signals including specific wavelengths or frequencies can be used to perform the communication network. The required information can be formed by the spatial soliton pulses generation using an analog to digital electronic convertor system [29, 30]. Furthermore, the applications such as quantum repeater, quantum entangled photon source are available, which can complete the concept of quantum optical communication networks [31].

3.4 Multi Users Optical Communication Applying Multi Solitons

Using the system including two microring resonators connected to an add/drop filter system, the multi soliton can be generated. From Fig. 3.11a the input soliton pulse has 20 ns pulse width, peak power of 500 mW. The ring radii are $R_1 = 10$, $R_2 = 5$, and $R_d = 200$ μm. The fixed parameter are selected to $\lambda_0 = 1.55$ μm, $n_0 = 3.34$ (IngaAsp/InP), $A_{eff} = 0.25$ μm^2, $\alpha = 0.5$ dB mm^{-1}, $\gamma = 0.1$. The coupling coefficients range from 0.50 to 0.975, where the nonlinear refractive index is $n_2 = 2.2 \times 10^{-17}$ m^2/W and the wave guided loss used is 0.5 dB mm^{-1}. Optical signals are sliced into smaller signals broadening over the band as shown in Fig. 3.11b. Therefore, large bandwidth signal is formed within the first ring device, where compress bandwidth with smaller group velocity is attained inside the ring R_2, such as filtering signals. Localized soliton pulses are formed, when resonant condition is performed, given in Fig. 3.11d. However, there are two types of temporal and spatial soliton pulses. Here, the multi soliton pulses with FSR and FHWM of 600 and 10 pm are generated.

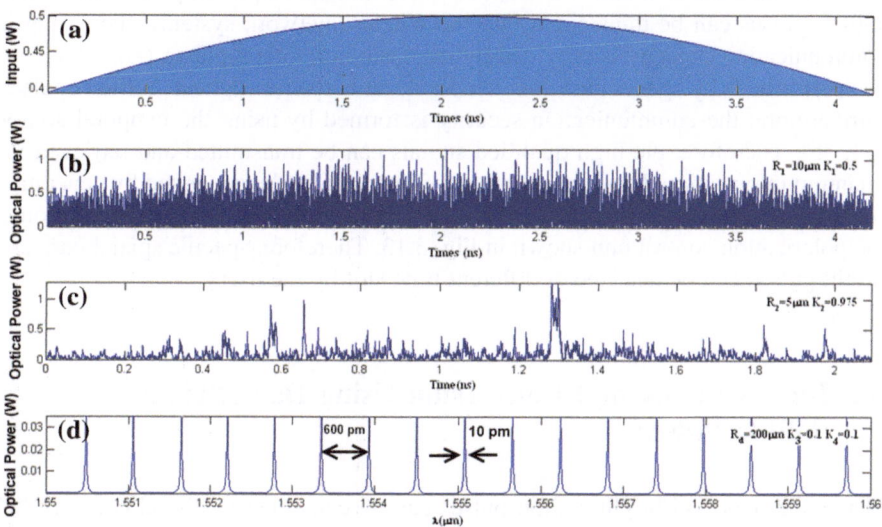

Fig. 3.11 Results of the multi-soliton pulse generation, **a** input soliton, **b** large bandwidth signals, **c** temporal soliton, **d** spatial soliton with FSR of 600 pm, and FWHM of 10 pm

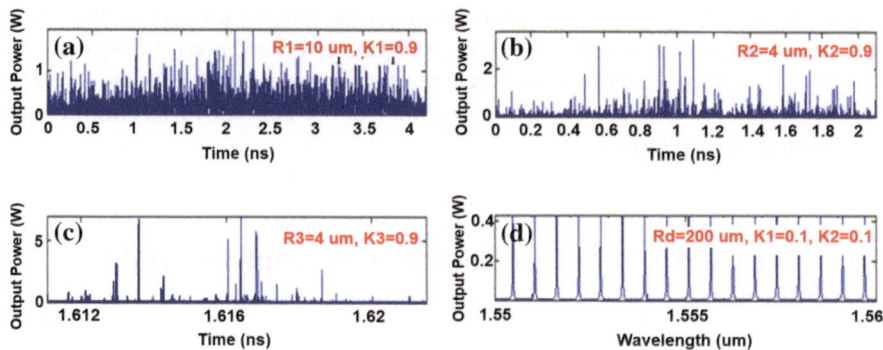

Fig. 3.12 Results of the multi-soliton pulse generation where **a** the large bandwidth signals, **b** and **c** temporal solitons, **d** the localized spatial solitons

Figure 3.12 show the generation of optical multi soliton signals, where $R_1 = 10$ μm, $R_2 = R_3 = 4$ μm, and $R_d = 200$ μm. Therefore, by using suitable ring resonator parameters, localized wavelength can be obtained. From Fig. 3.12a large bandwidth signals, **b** and **c** trapping of temporal soliton pulses, **d** localized spatial multi solitons. Amplification of optical soliton is performed used to long transmission link. The power distribution of the output pulses can be executed via the add/drop filter with radius of R_d as shown in Fig. 3.12d.

Therefore, the proposed system is suitable for the multi soliton pulses generation, which is available for high performance network. Since, an optical soliton communication has been realized as a good candidate for long distance communication, therefore, increase of soliton wavelengths is recommended [32, 33]. Generated multi soliton pulses can be transmitted into the wireless network systems. Increasing in communication capacity is provided by increasing of soliton pulses (λ_i), which can be performed by generation of temporal and spatial optical soliton pulses. Furthermore, the communication security is formed by using the temporal soliton [34, 35]. Therefore, the high qualified signals can be transmitted and retrieved via quantum information. Here the quantum information can be generated by generation of dark and bright optical solitons when the optical multi soliton pulses pass through the polarization control unit shown in Fig. 3.13. Therefore, specific spatial dark and bright pulses can be detected at different time slot by the users.

3.5 Binary to Decimal Conversion Using Dark/Bright Soliton Pulses

Generated dark and bright soliton pulses can be converted to digital codes of "0" and "1" by using analog to digital electronic convertor system [36, 37]. This system is known as optical binary to decimal convertor system which is applicable to

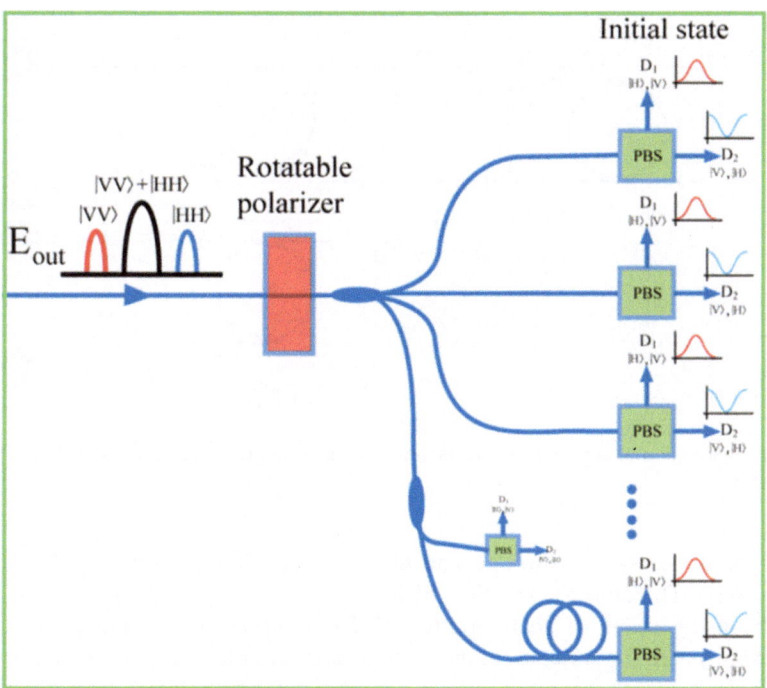

Fig. 3.13 System of dark and bright generation applicable for coding process, where PBS is a beam splitter and Ds are the detectors

generate data. Therefore, in operation, the large bandwidth within the MRR can be generated by using an optical soliton input into the device [38, 39]. The localized soliton pulse is generated, whereas the required signals included specific wavelengths can perform the communication network [40, 41]. The data can be formed by using the spatial soliton pulses. The proposed optical binary to decimal convertor system is show in Fig. 3.14. The input and control light pulse trains are input into the first add/drop optical filter (MRR1) using dark soliton (logic '0') or the bright soliton (logic '1'). First, the dark soliton is converted to be dark and bright soliton via the add/drop optical filter, which they can be seen at the through and drop ports with π phase shift. By using the add/drop optical filter (MRR2 and MRR3), both input signals are generated by the first stage add/drop optical filter. Next, the input data, "Y" with logic "0" (dark soliton) and logic "1" (bright soliton) are added into the both add ports, where the dark-bright soliton conversion with π phase shift is operated again. For large scale, results obtained are simultaneously seen by T2, D2, T3 and D3 at the drop and through ports for optical logic operation.

From Fig. 3.14, the optical pulse train X, Y is fed into MRR2 from input and add ports, respectively, in which the optical pulse trains that appear at the through and drop ports of MRR2 will be $X \bar{Y}$ and $X Y$ respectively. When the optical pulse train $\bar{X}$, $\bar{Y}$ is fed into MRR3 from input and add ports, respectively, the optical pulse

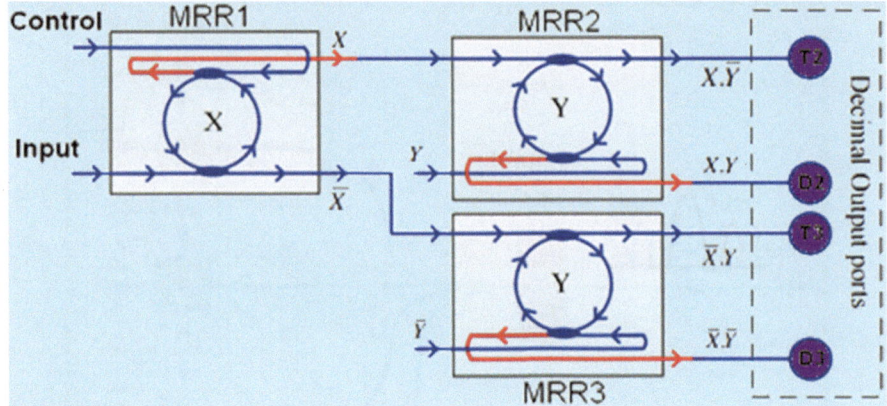

Fig. 3.14 System of binary to decimal convertor, where T and D are the through and drop ports of the system

trains that appear at the through and drop ports of MRR3 will be $\bar{X}Y$ and $\bar{X}\bar{Y}$, respectively. Therefore, generation of logic codes of "0" and "1", can be easily done by using series of beam splitters (B.S) connected to the binary to decimal convertor system. In simulation, the add/drop optical filter parameters are fixed for all coupling coefficients to be, $\kappa_s = 0.05$, $R_{ad} = 300\,\text{nm}$, $A_{eff} = 0.25\,\mu\text{m}^2$, $\alpha = 0.05\,\text{dB}\,\text{mm}^{-1}$. Here, the results show the generation of optical logic codes of "00", "01", "10" and "11", using the MRRs proposed system. A wireless router system can be used to decode the logic codes, transfer them via a wireless access point, and network communication system shown in Fig. 3.15.

Fig. 3.15 System of dark and bright generation applicable for coding process

A wireless access system transmits data to different users via wireless connection. The transmission of information can be sent to the Internet using a physical, wired Ethernet connection. This method also works in reverse, when the router system used to receive information from the Internet, translating it into an analog signal and sending it to the computer's wireless adapter.

3.6 Ultra-Short Multi Channel Soliton Pulse Generation

Gaussian beams with center wavelength of 1.55 μm and power of 600 mW are introduced into the add and input ports of the PANDA ring resonator. The result has been shown in Fig. 3.16. The linear and nonlinear refractive indices of the system are $n_0 = 3.34$ and $n_2 = 3.2 \times 10^{-17}$ respectively. In Fig. 3.16, the coupling coefficients of the PANDA ring resonator are given as $\kappa_0 = 0.2$, $\kappa_1 = 0.35$, $\kappa_2 = 0.1$ and $\kappa_3 = 0.95$, respectively and $\gamma = \gamma_1 = \gamma_2 = 0.1$. Here $R_{PANDA} = 300$ nm where $R_r = 180$ nm and $R_L = 200$ nm respectively. Figure 3.16a, b show the powers in the form of chaotic signals before entering the right ring of the PANDA system and amplification of signals during propagation of light inside right ring respectively, where Fig. 3.16c, d show the powers before entering the left ring and amplification of signals within the right ring respectively. We found that the signals are stable and seen within the system where the chaotic signals are generated at the through port shown in Fig. 3.16e.

In order to generate multi optical soliton, the chaotic signals from the PANDA ring resonator are input into the add/drop filter system. Figure 3.17a, b show the generation of multi soliton in the form of dark soliton and expansion of the through port signals respectively, where Fig. 3.17c, d represent multi soliton in the form of

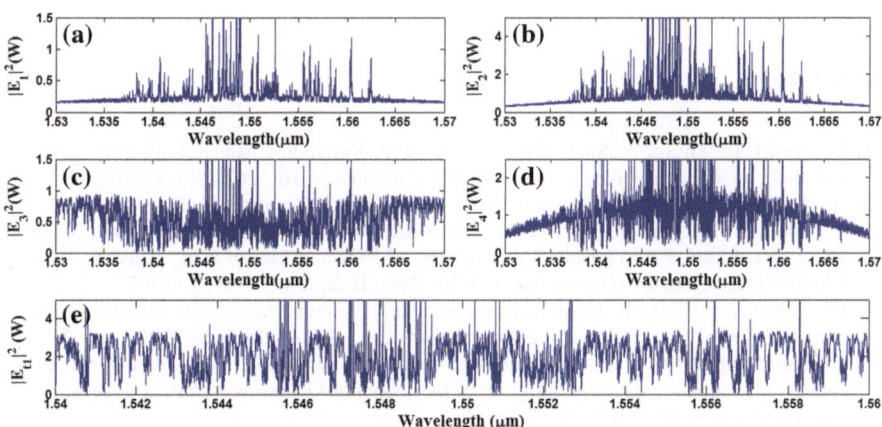

Fig. 3.16 Multi soliton signal generation using PANDA ring resonator system where **a, b, c** and **d** are powers inside the PANDA system and **e** is the output power from the throughput

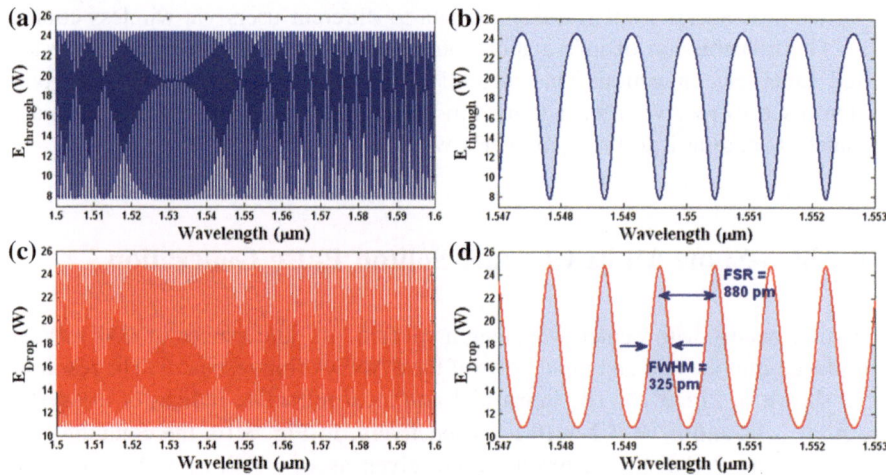

Fig. 3.17 Output multi soliton signal generation using an add/drop filter system, where **a** dark soliton at through port, **b** expansion of multi dark soliton, **c** bright soliton at drop port, and **d** expansion of multi bright soliton with FWHM and FSR of 325 and 880 pm respectively

bright solitons and expansion of the drop port signals respectively. The coupling coefficients of the add/drop filter system are given as $\kappa_4 = 0.9$, $\kappa_5 = 0.5$, where the radius of the ring is $R_{ad} = 130$ μm.

Generated multi optical soliton can be input into an optical receiver unit which is a quantum processing system used to generate high capacity packet of quantum information within the series of MRR's. In operation, the computing data can be modulated and input into the receiver unit which is encoded to the quantum signal processing system [42].

References

1. I.S. Amiri, R. Ahsan, A. Shahidinejad, J. Ali, P.P. Yupapin, Characterisation of bifurcation and chaos in silicon microring resonator. IET Commun. **6**(16), 2671–2675 (2012)
2. A. Afroozeh, I.S. Amiri, K. Chaudhary, J. Ali, P.P. Yupapin, Analysis of optical ring resonator, in *Advances in Laser and Optics Research* (Nova Science, New York, 2014)
3. I.S. Amiri, J. Ali, P.P. Yupapin, Enhancement of FSR and finesse using add/drop filter and PANDA ring resonator systems. Int. J. Mod. Phys. B **26**(04), 1250034 (2012)
4. I.S. Amiri, A. Afroozeh, Integrated ring resonator systems, in *Ring Resonator Systems to Perform Optical Communication Enhancement Using Soliton* (Springer, 2015), pp. 37–47
5. I.S. Amiri, S.E. Alavi, S.M. Idrus, Introduction of fiber waveguide and soliton signals used to enhance the communication security, in *Soliton Coding for Secured Optical Communication Link* (Springer, USA, 2015), pp. 1–16
6. A. Afroozeh, I.S. Amiri, A. Zeinalinezhad, *Micro Ring Resonators and Applications* (LAP Lambert Academic Publishing, Saarbrücken, 2014)

7. I.S. Amiri, A. Afroozeh, *Ring Resonator Systems to Perform the Optical Communication Enhancement Using Soliton* (Springer, USA, 2014)
8. S.E. Alavi, I.S. Amiri, S.M. Idrus, A.S.M. Supa'at, J. Ali, P.P. Yupapin, All optical OFDM generation for IEEE802.11a based on soliton carriers using microring resonators. IEEE Photonics J. **6**(1), 1–10 (2014)
9. I.S. Amiri, A. Nikoukar, J. Ali, GHz frequency band soliton generation using integrated ring resonator for WiMAX optical communication. Opt. Quant. Electron. **46**(9), 1165–1177 (2014)
10. I.S. Amiri, S.E. Alavi, S.M. Idrus, A.S.M. Supa'at, J. Ali, P.P. Yupapin, W-Band OFDM transmission for radio-over-fiber link using solitonic millimeter wave generated by MRR. IEEE J. Quantum Electron. **50**(8), 622–628 (2014)
11. I.S. Amiri, M.H. Khanmirzaei, M. Kouhnavard, P.P. Yupapin, J. Ali, Quantum entanglement using multi dark soliton correlation for multivariable quantum router, in ed. by A.M. Moran *Quantum Entanglement* (Nova Science Publisher, New York, 2012), pp. 111–122
12. I.S. Amiri, S.E. Alavi, S.M. Idrus, *Soliton Coding for Secured Optical Communication Link* (Springer, USA, 2014)
13. I.S. Amiri, J. Ali, Data signal processing via a manchester coding-decoding method using chaotic signals generated by a PANDA ring resonator. Chin. Opt. Lett. **11**(4), 041901(4) (2013)
14. I.S. Amiri, FWHM measurement of localized optical soliton, in *The International Conference for Nano materials Synthesis and Characterization* (Malaysia, 2011)
15. I.S. Amiri, S.E. Alavi, J. Ali, High-capacity soliton transmission for indoor and outdoor communications using integrated ring resonators. Int. J. Commun. Syst. **28**(1), 147–160 (2015)
16. I.S. Amiri, J. Ali, Optical quantum generation and transmission of 57–61 GHz frequency band using an optical fiber optics. J. Comput. Theor. Nanosci. **11**(10), 2130–2135 (2014)
17. I.S. Amiri, S. Soltanmohammadi, A. Shahidinejad, J. Ali, Optical quantum transmitter with finesse of 30 at 800-nm central wavelength using microring resonators. Opt. Quant. Electron. **45**(10), 1095–1105 (2013)
18. I.S. Amiri, J. Ali, Picosecond soliton pulse generation using a PANDA system for solar cells fabrication. J. Comput. Theor. Nanosci. **11**(3), 693–701 (2014)
19. I.S. Amiri, Optical soliton trapping for quantum key generation, in *The International Conference for Nano materials Synthesis and Characterization* (Malaysia, 2011)
20. I.S. Amiri, A. Nikoukar, S. E. Alavi, Soliton and radio over fiber (RoF) applications. (LAP LAMBERT Academic Publishing, Saarbrücken, 2014)
21. I.S. Amiri, S.E. Alavi, S.M. Idrus, Solitonic pulse generation and characterization by integrated ring resonators, *Presented at the 5th International Conference on Photonics 2014 (ICP2014)* (Kuala Lumpur, 2014)
22. I.S. Amiri, S.E. Alavi, S.M. Idrus, M. Kouhnavard, *Microring Resonator For Secured Optical Communication* (Amazon, USA, 2014)
23. I.S. Amiri, A. Afroozeh, Soliton generation based optical communication, in *Ring Resonator Systems to Perform Optical Communication Enhancement Using Soliton* (Springer, 2015), pp. 49–68
24. I.S. Amiri, M. Nikmaram, A. Shahidinejad, J. Ali, Generation of potential wells used for quantum codes transmission via a TDMA network communication system. Secur. Commun. Netw. **6**(11), 1301–1309 (2013)
25. I.S. Amiri, S.E. Alavi, S.M. Idrus, A. Afroozeh, J. Ali, *Soliton Generation by Ring Resonator for Optical Communication Application* (Nova Science Publishers, Hauppauge, 2014)
26. I.S. Amiri, A. Afroozeh, Mathematics of soliton transmission in optical fiber, in *Ring Resonator Systems to Perform Optical Communication Enhancement Using Soliton* (Springer, 2015), pp. 9–35
27. I.S. Amiri, S.E. Alavi, S.M. Idrus, Results of digital soliton pulse generation and transmission using microring resonators, in *Soliton Coding for Secured Optical Communication Link* (Springer, USA, 2015), pp. 41–56

28. I.S. Amiri, M. Ebrahimi, A.H. Yazdavar, S. Gorbani, S.E. Alavi, S.M. Idrus, J. Ali, Transmission of data with orthogonal frequency division multiplexing technique for communication networks using GHz frequency band soliton carrier. IET Commun. **8**(8), 1364–1373 (2014)
29. I.S. Amiri, A. Afroozeh, I.N. Nawi, M.A. Jalil, A. Mohamad, J. Ali, P.P. Yupapin, Dark soliton array for communication security. Procedia Eng. **8**, 417–422 (2011)
30. I.S. Amiri, A. Afroozeh, Spatial and temporal soliton pulse generation by transmission of chaotic signals using fiber optic link, in *Advances in Laser and Optics Research*, vol. 11 (Nova Science Publisher, New York, 2014)
31. I.S. Amiri, J. Ali, Nano particle trapping by ultra-short tweezer and wells using MRR interferometer system for spectroscopy application. Nanosci. Nanotechnol. Lett. **5**(8), 850–856 (2013)
32. I.S. Amiri, J. Ali, Generating highly dark-bright solitons by gaussian beam propagation in a PANDA Ring resonator. J. Comput. Theor. Nanosci. **11**(4), 1092–1099 (2014)
33. I.S. Amiri, A. Afroozeh, Introduction of soliton generation, in *Ring Resonator Systems to Perform Optical Communication Enhancement Using Soliton* (Springer, 2015), pp. 1–7
34. I.S. Amiri, K. Raman, A. Afroozeh, M.A. Jalil, I.N. Nawi, J. Ali, P.P. Yupapin, Generation of DSA for security application. Procedia Eng. **8**, 360–365 (2011)
35. I.S. Amiri, P. Naraei, J. Ali, Review and theory of optical soliton generation used to improve the security and high capacity of MRR and NRR passive systems. J. Comput. Theor. Nanosci. **11**(9), 1875–1886 (2014)
36. I.S. Amiri, A. Afroozeh, M. Bahadoran, Simulation and analysis of multisoliton generation using a PANDA ring resonator system. Chin. Phys. Lett. **28**(10), 104205 (2011)
37. I.S. Amiri, S.E. Alavi, S.M. Idrus, Theoretical background of microring resonator systems and soliton communication, in *Soliton Coding for Secured Optical Communication Link* (Springer, USA, 2015), pp. 17–39
38. P. Sanati, A. Afroozeh, I.S. Amiri, J. Ali, L.S. Chua, Femtosecond pulse generation using microring resonators for eye nano surgery. Nanosci. Nanotechnol. Lett. **6**(3), 221–226 (2014)
39. I.S. Amiri, B. Barati, P. Sanati, A. Hosseinnia, H.R. Mansouri Khosravi, S. Pourmehdi, A. Emami, J. Ali, Optical stretcher of biological cells using sub-nanometer optical tweezers generated by an add/drop microring resonator system. Nanosci. Nanotechnol. Lett. **6**(2), 111–117 (2014)
40. S.E. Alavi, I.S. Amiri, S.M. Idrus, A.S.M. Supa'at, Generation and wired/wireless transmission of IEEE802.16 m signal using solitons generated by microring resonator. Opt. Quant. Electron. (2014)
41. I.S. Amiri, S.E. Alavi, S.M. Idrus, A. Nikoukar, J. Ali, IEEE 802.15.3c WPAN standard using millimeter optical soliton pulse generated by a panda ring resonator. IEEE Photonics J. **5**(5), 7901912 (2013)
42. I.S. Amiri, J. Ali, Femtosecond optical quantum memory generation using optical bright soliton. J. Comput. Theor. Nanosci. **11**(6), 1480–1485 (2014)

Chapter 4
Ultra-Short Solitonic Pulses Used in Optical Communication

Abstract Microring resonators (MRRs) have shown great promise for application in many research areas such as optical communications in the micro- and nanoscale regime. An optical soliton is a powerful laser pulse that can be used to enlarge the optical bandwidth during propagation within a nonlinear MRR. The advantage of the proposed system is that the transmitter can be fabricated on-chip or operated alternatively by a single device. The MRR's performance can be described in terms of several parameters such as the free spectrum range (FSR), full width at half maximum (FWHM), and finesse. MRR systems, provides the foundation for the development of new transmission techniques. Traveling of light inside the proposed ring system is analyzed by manipulating of nonlinear refractive index, coupling coefficient and the radius of the ring resonator. Therefore, controlling the round trip times and the input power of the system can be used to deal with and control the bifurcation and chaotic signals, where it is used in many applications in photonics communication such as signal processing or digital implementations. Ultra-short high quality soliton signals can be performed and transmit via a network system. Here, the optical information can be generated by large bandwidth of the arbitrary wavelength and frequency of light pulse signals via a series of MRRs device. Optical communication capacity can be increased by the multi soliton pulses generation. These type of signals are used widely to generate quantum data applicable to quantum network communication.

Keyword Microring resonator (MRR) · Micro- and nanoscale regime · Transmission techniques · Ultra-short high quality soliton signals · Multi soliton pulses

Microring resonators (MRRs) have shown great promise for applications in many research areas such as computers, communications, and signal processing in the micro- and nanoscale regime [1–3]. MRRs have attracted interest in recent years owing to their versatile applications such as optical filters, optical sensors, optical transmitters, Wavelength Division Multiplexing (WDM), and on-off switches [4]. An optical soliton is a powerful laser pulse that can be used to enlarge the optical bandwidth during propagation within a nonlinear MRR [5].

© The Author(s) 2015
I. Sadegh Amiri and H. Ahmad, *Optical Soliton Communication Using Ultra-Short Pulses*, SpringerBriefs in Applied Sciences and Technology, DOI 10.1007/978-981-287-558-7_4

By using the proposed system, the transceiver can be integrated and operated using a single device [6]. One important aspect of the system is that the required soliton pulses with specific key parameters such as finesse can be obtained at the drop/through ports of the system by tuning the parameters of the system [7, 8]. The advantage of the proposed system is that the transmitter can be fabricated on-chip or operated alternatively by a single device [9]. Although, laser sources presently face limited market acceptance owing to shortcomings such as a limited tunability range and cumbersome size, these constraints are rapidly disappearing with the invention of new classes of tunable diode laser setups [10, 11]. The main advantages of these new lasers include reliability, compactness, and efficiency.

To maintain soliton pulse propagation through the ring resonator, suitable coupling power is required. The MRR's performance can be described in terms of several parameters such as the free spectrum range (FSR), full width at half maximum (FWHM), and finesse [12]. Applications to communication systems require high performance, low loss, high speed, low cost, and simplicity in terms of both fabrication and setup. Optical channel filters with low insertion loss, wide FSR (high selectivity), and high stop band rejection are required for applications such as dense wavelength multiplexing (DWDM) [13, 14]. The fractional delay for input soliton pulses is a suitable measure of the number of pulse widths in which a pulse can be delayed. Therefore, to maximize the fractional delay for the input soliton pulse, the resonator bandwidth should be selected properly. This still allows for selection between a small, high-finesse resonator or a larger and proportionally lower-finesse resonator. If both tolerate the same loss per round trip prescribed by splices and coupling losses, the lower finesse device that approximates a phase-only device has a more uniform transmission across the FSR.

Many theoretical and experimental designs have been proposed to optimize the filter response and other properties using various coupling coefficients and radii [15]. Although the soliton interaction is elastic, the interaction between solitons would affect DWDM. However, this problem can be resolved by designing a suitable system with desirable free spectrum arrangement. Exciting new technological progress, particularly in the field of tunable narrow band laser systems [16], multiple transmission, and MRR systems, provides the foundation for the development of new transmission techniques. In addition to improvements in efficiency and beam quality, these soliton sources provide better sensitivity required for sensing systems, leading to improved process efficiencies and new fields of optical sensors [17–19]. The soliton pulses are so stable that the shape and velocity are preserved while travelling along the medium [20, 21].

We have presented nonlinear effects of the single ring resonator as optical bifurcation and chaos [22, 23]. Traveling of light inside the proposed ring system is analyzed by manipulating of nonlinear refractive index, coupling coefficient and the radius of the ring resonator [24, 25]. Results have shown that the bifurcation and chaos can be occurred in different roundtrip times and input power. Occurrence of bifurcation at lower input power or smaller round trip is a beneficial effect in order to improve the nonlinear microring system [26]. Therefore, controlling the round trip times and the input power of the system can be used to deal with and control the bifurcation and chaotic signals, where it is used in many applications in photonics communication such as signal processing or digital implementations [27]. In this book generation of quantum information was performed using chaotic signals which are transmitted via wireless networks and information transmission system [28, 29].

As conclusion, a system of MRRs was presented in which, high quality soliton signals can be performed and transmit via a network system [30]. Optical input pulse of a dark soliton or a Gaussian beam is introduced into the nonlinear Kerr-type MRR devices. MRR system is connected to a rotatable polarizer and a beam splitter, where this system can be used to generate localized optical soliton pulses applicable in public network systems. Here, the optical information can be generated by large bandwidth of the arbitrary wavelength and frequency of light pulse signals via a series of MRRs device. Spatial optical soliton pulse can be used to perform the networks communication [31]. In this book, localized optical pulse with frequencies of 10.7 and 16 MHz and wavelengths of 206.9, 1448, 2169 and 2489 nm are simulated. Further, more frequency and wavelength bands are available for many applications in networks communication [32, 33].

We proposed interesting concept of information generation where the system of micro or nano ring resonator are used to generate high capacity of multi optical soliton, connected to a beam splitter and analog to digital convertor system. Chaotic signal generation using a soliton pulse in the nonlinear MRRs is presented [34]. Optical communication capacity can be increased by the multi soliton pulses generation, where more soliton channels can be generated by using the MRR system. The required channels are obtained by filtering the large bandwidth signals using an add/drop filter system [35]. The proposed system consists of a series of MRR devices, where the digital codes of "0" and "1" can be generated within the optical binary to decimal convertor system. The advantage of the system is that the clear signal can be retrieved by the specific add/drop filter. Generated data can be transmitted into network communication systems via a wireless system.

Extensive pulses in the form of chaotic signals can be generated by using a PANDA ring resonator system. This system is connected to an add/drop filter system in order to generate highly multi optical soliton. Gaussian beams with center wavelength of 1.55 μm, are introduced into the input and add ports of the PANDA system which are sufficient to generate high capacity of soliton. Therefore, interior signals of the PANDA system can be controlled and tuned [36]. Generated chaotic signals from the PANDA system can be input into the add/drop filter system. The add/drop system will filter the chaotic signals in which the multi pulses of soliton with FWHM and FSR of 325 and 880 pm can be generated. These types of signals are used widely to generate quantum data applicable to quantum network communication. Therefore, we have proposed an interesting concept of internet communication based on quantum information where the use of data encoding for high capacity communication via optical network link is plausible. The advantage of the microring resonator device is that the quantum data patterns can be randomly generated by changing the setting parameters.

References

1. A. Afroozeh, I.S. Amiri, K. Chaudhary, J. Ali, P.P. Yupapin, Analysis of optical ring resonator, in *Advances in Laser and Optics Research* (Nova Science, New York, 2014)
2. I.S. Amiri, S.E. Alavi, S.M. Idrus, Introduction of fiber waveguide and soliton signals used to enhance the communication security, in *Soliton Coding for Secured Optical Communication Link* (Springer, 2015), pp. 1–16
3. I.S. Amiri, A. Nikoukar, S.E. Alavi, *Soliton and Radio over Fiber (RoF) Applications* (LAP Lambert Academic Publishing, Saarbrücken, 2014)
4. I.S. Amiri, J. Ali, Generating highly dark-bright solitons by gaussian beam propagation in a PANDA ring resonator. J. Comput. Theor. Nanosci. **11**(4), 1092–1099 (2014)
5. I.S. Amiri, A. Afroozeh, Mathematics of soliton transmission in optical fiber, in *Ring Resonator Systems to Perform Optical Communication Enhancement Using Soliton* (Springer, 2015), pp. 9–35
6. S.E. Alavi, I.S. Amiri, S.M. Idrus, A.S.M. Supa'at, J. Ali, P.P. Yupapin, All optical OFDM generation for IEEE802.11a based on soliton carriers using microring resonators. IEEE Photonics J. **6**(1), 1–10 (2014)
7. I.S. Amiri, J. Ali, P.P. Yupapin, Enhancement of FSR and finesse using add/drop filter and PANDA ring resonator systems. Int. J. Mod. Phys. B **26**(04), 1250034 (2012)
8. I.S. Amiri, S. Soltanmohammadi, A. Shahidinejad, J. Ali, Optical quantum transmitter with finesse of 30 at 800-nm central wavelength using microring resonators. Opt. Quant. Electron. **45**(10), 1095–1105 (2013)
9. S.E. Alavi, I.S. Amiri, S.M. Idrus, A.S.M. Supa'at, Generation and wired/wireless transmission of IEEE802.16 m signal using solitons generated by microring resonator. Opt. Quant. Electron. (2014)
10. I.S. Amiri, A. Afroozeh, Integrated ring resonator systems, in *Ring Resonator Systems to Perform Optical Communication Enhancement Using Soliton* (Springer, 2015), pp. 37–47
11. I.S. Amiri, S.E. Alavi, S.M. Idrus, Results of digital soliton pulse generation and transmission using microring resonators, in *Soliton Coding for Secured Optical Communication Link* (Springer, 2015), pp. 41–56
12. I.S. Amiri, A. Afroozeh, *Ring Resonator Systems to Perform Optical Communication Enhancement Using Soliton* (Springer, 2014)
13. I.S. Amiri, S.E. Alavi, S.M. Idrus, *Soliton Coding for Secured Optical Communication Link* (Springer, 2014)
14. I.S. Amiri, A. Afroozeh, Soliton generation based optical communication, in *Ring Resonator Systems to Perform Optical Communication Enhancement Using Soliton* (Springer, 2015), pp. 49–68
15. I.S. Amiri, A. Afroozeh, Introduction of soliton generation, in *Ring Resonator Systems to Perform Optical Communication Enhancement Using Soliton* (Springer, 2015), pp. 1–7
16. P. Sanati, A. Afroozeh, I.S. Amiri, J. Ali, L.S. Chua, Femtosecond pulse generation using microring resonators for eye nano surgery. Nanosci. Nanotechnol. Lett. **6**(3), 221–226 (2014)
17. I.S. Amiri, J. Ali, Femtosecond optical quantum memory generation using optical bright soliton. J. Comput. Theor. Nanosci. **11**(6), 1480–1485 (2014)
18. I.S. Amiri, J. Ali, Nano particle trapping by ultra-short tweezer and wells using MRR interferometer system for spectroscopy application. Nanosci. Nanotechnol. Lett. **5**(8), 850–856 (2013)
19. I.S. Amiri, B. Barati, P. Sanati, A. Hosseinnia, H.R. Mansouri Khosravi, S. Pourmehdi, A. Emami, J. Ali, Optical stretcher of biological cells using sub-nanometer optical tweezers generated by an add/drop microring resonator system. Nanosci. Nanotechnol. Lett. **6**(2), 111–117 (2014)
20. I.S. Amiri, A. Afroozeh, M. Bahadoran, Simulation and analysis of multisoliton generation using a PANDA ring resonator system. Chin. Phys. Lett. **28**(10), 104205 (2011)

21. I.S. Amiri, A. Afroozeh, Spatial and temporal soliton pulse generation by transmission of chaotic signals using fiber optic link, in *Advances in Laser and Optics Research*, vol. 11 (Nova Science Publisher, New York, 2014)

22. I.S. Amiri, R. Ahsan, A. Shahidinejad, J. Ali, P.P. Yupapin, Characterisation of bifurcation and chaos in silicon microring resonator. IET Commun. **6**(16), 2671–2675 (2012)

23. I.S. Amiri, S.E. Alavi, S.M. Idrus, A. Afroozeh, J. Ali, *Soliton Generation by Ring Resonator for Optical Communication Application* (Nova Science Publishers, Hauppauge, 2014)

24. I.S. Amiri, J. Ali, Data signal processing via a manchester coding-decoding method using chaotic signals generated by a PANDA ring resonator. Chin. Opt. Lett. **11**(4), 041901(4) (2013)

25. I.S. Amiri, S.E. Alavi, S.M. Idrus, Theoretical background of microring resonator systems and soliton communication, in *Soliton Coding for Secured Optical Communication Link* (Springer, 2015), pp. 17–39

26. I.S. Amiri, M. Ebrahimi, A.H. Yazdavar, S. Gorbani, S.E. Alavi, S.M. Idrus, J. Ali, Transmission of data with orthogonal frequency division multiplexing technique for communication networks using GHz frequency band soliton carrier. IET Commun. **8**(8), 1364–1373 (2014)

27. I.S. Amiri, S.E. Alavi, S.M. Idrus, A. Nikoukar, J. Ali, IEEE 802.15.3c WPAN standard using millimeter optical soliton pulse generated by a panda ring resonator. IEEE Photonics J. **5**(5), 7901912 (2013)

28. I.S. Amiri, J. Ali, Optical quantum generation and transmission of 57–61 GHz frequency band using an optical fiber optics. J. Comput. Theor. Nanosci. **11**(10), 2130–2135 (2014)

29. I.S. Amiri, M.H. Khanmirzaei, M. Kouhnavard, P.P. Yupapin, J. Ali, Quantum entanglement using multi dark soliton correlation for multivariable quantum router, in ed. by A.M. Moran *Quantum Entanglement* (Nova Science Publisher, New York, 2012), pp. 111–122

30. I.S. Amiri, S.E. Alavi, J. Ali, High-capacity soliton transmission for indoor and outdoor communications using integrated ring resonators. Int. J. Commun. Syst. **28**(1), 147–160 (2015)

31. I.S. Amiri, S.E. Alavi, S.M. Idrus, Solitonic pulse generation and characterization by integrated ring resonators. *Presented at the 5th International Conference on Photonics 2014 (ICP2014)*, Kuala Lumpur (2014)

32. I.S. Amiri, A. Nikoukar, J. Ali, GHz frequency band soliton generation using integrated ring resonator for WiMAX optical communication. Opt. Quant. Electron. **46**(9), 1165–1177 (2014)

33. I.S. Amiri, S.E. Alavi, S.M. Idrus, A.S.M. Supa'at, J. Ali, P.P. Yupapin, W-Band OFDM transmission for radio-over-fiber link using solitonic millimeter wave generated by MRR. IEEE J. Quantum Electron. **50**(8), 622–628 (2014)

34. I.S. Amiri, P. Naraei, J. Ali, Review and theory of optical soliton generation used to improve the security and high capacity of MRR and NRR passive systems. J. Comput. Theor. Nanosci. **11**(9), 1875–1886 (2014)

35. I.S. Amiri, M. Nikmaram, A. Shahidinejad, J. Ali, Generation of potential wells used for quantum codes transmission via a TDMA network communication system. Secur. Commun. Netw. **6**(11), 1301–1309 (2013)

36. I.S. Amiri, J. Ali, Picosecond soliton pulse generation using a PANDA system for solar cells fabrication. J. Comput. Theor. Nanosci. **11**(3), 693–701 (2014)